STUDENT SOLUTIONS MANUAL

Cynthia G. Zoski
Department of Chemistry
The University of Texas at Austin
The Center for Electrochemistry
Austin, TX USA

Johna Leddy
Department of Chemistry
The University of Iowa
Iowa City, IA USA

to accompany

ELECTROCHEMICAL METHODS

Fundamentals and Applications

Third Edition

T0338008

STUDENT SOLUTIONS MANUAL

Cynthia G. Zoski
Department of Chemistry
The University of Texas at Austin
The Center for Electrochemistry
Austin, TX USA

Johna Leddy
Department of Chemistry
The University of Iowa
Iowa City, IA USA

to accompany

ELECTROCHEMICAL METHODS

Fundamentals and Applications

Third Edition

Allen J. Bard
Larry R. Faulkner
Department of Chemistry and Center for Electrochemistry,
University of Texas at Austin

Henry S.White
Department of Chemistry, University of Utah

WILEY

Registered Offices
John Wiley & Sons, Inc., 111 River Street, Hoboken, NJ 07030, USA
John Wiley & Sons Ltd, The Atrium, Southern Gate, Chichester, West Sussex, PO19 8SQ, UK

For details of our global editorial offices, customer services, and more information about Wiley products visit us at www.wiley.com.

Wiley also publishes its books in a variety of electronic formats and by print-on-demand. Some content that appears in standard print versions of this book may not be available in other formats.

Library of Congress Cataloging-in-Publication Data applied for:

Paperback ISBN: 9781119524069

Cover Design: Wiley
Cover Image: Courtesy of Allen J. Bard, Larry R. Faulkner and Henry S. White

SKY10086688_100824

TABLE OF CONTENTS

PREFACE

This 3rd edition Solutions Manual is a supplement to *Electrochemical Methods: Fundamentals and Applications* (3rd Edition, 2022), authored by Allen J. Bard, Larry R. Faulkner, and Henry S. White. We hope the problem solutions are of use to practitioners and students of electrochemistry.

We extend sincere thanks to Al Bard, Larry Faulkner, and Henry White for many enlightening discussions and to Larry and Henry for their dedicated assistance in solving problems, proofreading, and drawing figures to support solution to problems in this edition of the Manual. We are grateful to D.C. Dunwoody whose TEX formatting in the 2nd edition Solutions Manual is applied to this 3rd edition. We also extend thanks to Editors Sarah Higginbotham and Vishali Chandra Mohan at Wiley for facilitating this project.

We welcome comments and queries on our approach to the problems. These can be addressed to Cynthia G. Zoski (zoskicg@gmail.com) and Johna Leddy (johna-leddy@uiowa.edu). Additional information and updates may also be found at www.wiley.com/go/solutionmanual/electrochemicalmethods3e.

Al Bard passed away on February 11, 2024. Al was a modest man and eminent electrochemist, who looked on his academic and research career as "the most fun of anything to do". Al loved doing science, the challenge of it, and working with new people and young people. Al was most proud of his 360 graduate students and postdocs, and their students, and their students after them, as they continue building the edifice of science. This is Al's legacy. With his students and postdocs as co-authors, Al's peer-reviewed publications total 1011. Al considered the 3rd Edition of *Electrochemical Methods* a significant accomplishment and contribution to the society of electrochemists. He continued working on the 3rd Edition into his retirement years and until its publication in 2022. The 3rd Edition updates the previous two editions with the most recent electrochemical advances. Al's command of electrochemistry and his ability to convey the science are unparalleled. Al was a mentor, a colleague, and a dear friend. We will miss him. We are pleased to dedicate this 3rd edition Solutions Manual to Al.

Cynthia G. Zoski and Johna Leddy

1 OVERVIEW OF ELECTRODE PROCESSES

Problem 1.1$^{\copyright}$ **(a).** In approaching this kind of problem, it is useful to list all the couples in Table C.1 that are relevant to the system.

E^0 vs. NHE (V)	Reaction
1.229	$O_2 + 4H^+ + 4e \rightleftharpoons H_2O$
1.188	$Pt^{2+} + 2e \rightleftharpoons Pt$
0.340	$Cu^{2+} + 2e \rightleftharpoons Cu$
0.159	$Cu^{2+} + e \rightleftharpoons Cu^+$
0.000	$2H^+ + 2e \rightleftharpoons H_2$
-0.4025	$Cd^{2+} + 2e \rightleftharpoons Cd$

Alternatively, a graphical representation may prove useful. Here, the standard or formal potentials for each redox couple are plotted on a potential axis. The species present in solution are underlined. Note the reduced half of the couple is noted toward more negative potentials. The vertical line indicates the approximate potential range where both halves of the redox couple can exist. For electrode potentials positive of a given line, the oxidized half of the couple is stable at the electrode surface; for electrode potentials negative of the line, the reduced form is stable. Note that for $n = 1$, electrode potentials within 118 mV of E^0 require no less than 1% of either the oxidized or reduced halves of the couple as given by $\log \frac{[O]}{[R]} = -n\left(E - E^0\right)/0.059$.

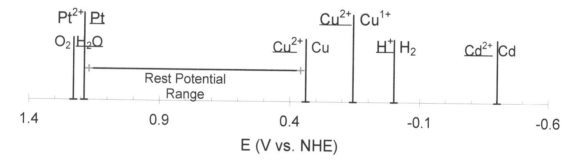

The composition of the system dictates that the rest (zero current) potential be more positive than $E^0_{Cu^{2+}/Cu}$ and more negative than $E^0_{O_2/H_2O}$ or $E^0_{Pt^{2+}/Pt}$, i.e., between about 0.34 V and 1.2 V vs. NHE. Graphically, this is apparent because this is the voltage range over which the oxidized (Cu^{2+}) and reduced species (Pt or H_2O) present in the solution are most adjacent on the graph. This defines a zone of stability set by the oxidized and reduced species. (Note that the cell would not be at equilibrium if oxidized and reduced species of two or more couples were present such that they were on the outer sides of the lines. For example, if the solution contained Cu and O_2, there would be a thermodynamic driving force for these species to react spontaneously to form water and Cu^{2+}.) In the rest potential range, the potential is not well defined in a thermodynamic sense; the electrode is not well poised, because no couple has both oxidized and reduced forms present. Calculation of the equilibrium potential by the Nernst equation cannot be made.

Electrochemical Methods: Fundamentals and Applications, Third Edition, Student Solutions Manual. Cynthia G. Zoski and Johna Leddy.
© 2025 John Wiley & Sons Ltd. Published 2025 by John Wiley & Sons Ltd.

Chapter 1   OVERVIEW OF ELECTRODE PROCESSES

Current will flow when the potential is moved negatively from the rest potential to about 0.340 V (or $0.340 + (-0.2412) = 0.099$ V vs. SCE) so that Cu^{2+} is reduced at the electrode surface first.

$$Cu^{2+} + 2e \rightleftharpoons Cu \qquad \text{(first reduction, } \approx 0.1 \text{ V vs. SCE)}$$

A positive movement from the rest potential first causes significant current flow when platinum and water are oxidized.

$$Pt \rightleftharpoons Pt^{2+} + 2e \qquad \text{(first oxidations, } \approx 1.0 \text{ V vs. SCE)}$$
$$2H_2O \rightleftharpoons O_2 + 4H^+ + 4e$$

Actually, Pt would form a thin oxide film, then it would stabilize, and only the oxygen evolution reaction would occur, with a significant negative (anodic) current flow due to the oxidation of water, which marks the positive background limit. The current-potential curve would look like the following with a transition from a limiting positive (cathodic) current flow due to Cu^{2+} reduction to Cu metal that plates onto the Pt UME to a significant cathodic current beginning at -0.1 V vs SCE due to H^+ (2M in solution) reduction,which marks the negative background limit. Although Cd^{2+} is present in the solution, its reduction to Cd cannot be observed, because this reduction occurs at about -0.6 V vs. SCE, far beyond the negative background limit.

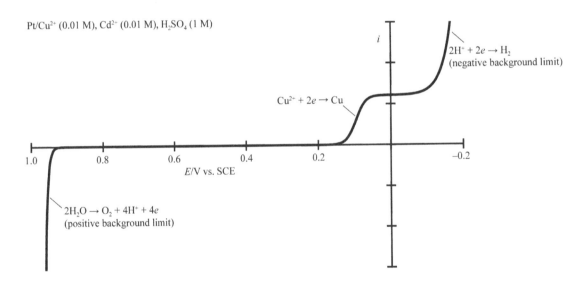

Pt/Cu^{2+} (0.01 M), Cd^{2+} (0.01 M), H_2SO_4 (1 M)

The reader may be puzzled regarding the locations of the background limits, which are less extreme than the related standard potentials. One must remember that a background curve represents just the foot of an enormous wave supported by a highly available electroreactant (in this case, H^+ or H_2O). The half-wave potential for that wave might be near the standard potential of the couple, but on the current scale that allows observation of voltammetric features of interest, one never approaches the standard potential before the background current becomes prohibitive. In general, a background limit based on a reversible or quasireversible couple is less extreme than the corresponding standard potential by about 0.1 V.

(b). The couples to be considered from Table C.1 are as follows:

E^0 vs. NHE (V)	Reaction
1.3583	$Cl_{2(g)} + 2e \rightleftharpoons 2Cl^-$
1.229	$O_2 + 4H^+ + 4e \rightleftharpoons H_2O$
1.188	$Pt^{2+} + 2e \rightleftharpoons Pt$
0.15	$Sn^{4+} + 2e \rightleftharpoons Sn^{2+}$
0.000	$2H^+ + 2e \rightleftharpoons H_2$
-0.1375	$Sn^{2+} + 2e \rightleftharpoons Sn$

The graphical representation is as follows.

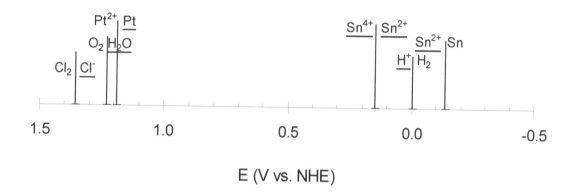

Because both Sn^{4+} and Sn^{2+} are present, the system is poised and a well-defined thermodynamic equilibrium potential exists. From the Nernst equation (1.3.13), the equilibrium potential is 0.15 V vs. NHE or -0.09 V vs. SCE. Moving negatively from this potential favors Sn^{2+} at the electrode and requires reduction of Sn^{4+}.

$$Sn^{4+} + 2e \rightleftharpoons Sn^{2+} \qquad \text{(first negative process, at} \approx \text{-0.09 V vs. SCE)}$$

Likewise, a move positive of the rest potential favors Sn^{4+} at the electrode and drives oxidation.

$$Sn^{2+} \rightleftharpoons Sn^{4+} + 2e \qquad \text{(first positive process, also at} \approx \text{-0.09 V vs. SCE)}$$

The current potential curve resembles the following where the equilibrium potential (-0.09 V vs SCE) for the Sn^{4+}/Sn^{2+} couple is shown because both tin cations are present in the electrochemical cell at the same concentration. Moving negative from this potential for Sn^{4+}/Sn^{2+} and positive from this potential for Sn^{2+}/Sn^{4+} leads to a cathodic and anodic limiting current respectively that are equal but opposite in sign. Moving yet more negative in potential from the cathodic limiting current leads to a significant increase in cathodic current at -0.1 V vs SCE due to proton (1 M) reduction, which corresponds to the negative background limit. The limiting current for reduction of Sn^{4+} is not fully resolved from the current rise at this limit. Moving more positive in potential from about +0.9 V from the anodic limiting current leads to a significant excursion of the anodic current due to water oxidation, which corresponds to the positive background limit.

Pt/Sn²⁺ (0.01 M), Sn⁴⁺ (0.01 M), HCl (1 M)

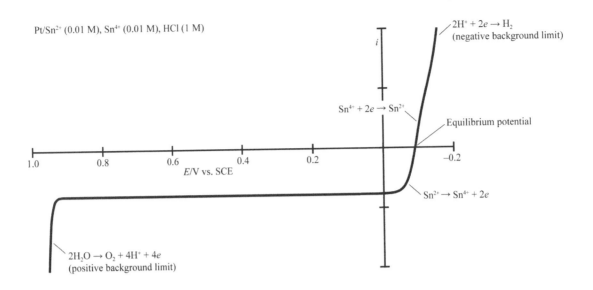

(c). The couples to be considered from Table C.1 are as follows:

E^0 vs. NHE (V)	Reaction
1.3583	$Cl_{2(g)} + 2e \rightleftharpoons 2Cl^-$
1.229	$O_2 + 4H^+ + 4e \rightleftharpoons H_2O$
0.7960	$Hg_2^{2+} + 2e \rightleftharpoons 2Hg$
0.26816	$Hg_2Cl_2 + 2e \rightleftharpoons 2Hg + 2Cl^-$
0.000	$2H^+ + 2e \rightleftharpoons H_2$
-0.3515	$Cd^{2+} + 2e \rightleftharpoons Cd(Hg)$
-0.7656	$Zn^{2+} + 2e \rightleftharpoons Zn(Hg)$

The graphical representation is shown.

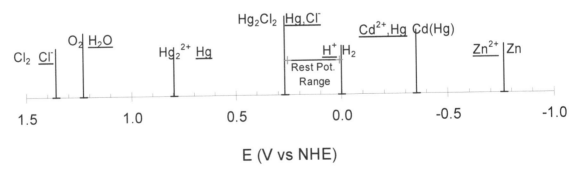

As in (a), the system is unpoised and the rest potential is not well defined, but exists between $E^0_{H^+/H^2} = 0.0$ V vs. NHE and $E^0_{Hg_2Cl_2/Hg}$ at 0.26816 V vs. NHE; that is, between -0.2412V and 0.02696 V vs. SCE. The first oxidation occurs when the potential is drawn to more positive values than ≈ -0.1 V vs. SCE, where the following reaction begins.

$$2Hg + 2Cl^- \rightleftharpoons Hg_2Cl_2 + 2e \qquad \text{(first positive process, } E^0 \approx 0.03 \text{ V vs. SCE)}$$

The chart and diagram predict that the first reduction will be the evolution of hydrogen.

$$2H^+ + 2e \rightleftharpoons H_2 \qquad \text{(predicted first reduction, } \approx \text{-0.24 V vs. SCE)}$$

However, this reaction is extremely slow on mercury (i.e., it has a high overpotential; see Section 1.1.7(c) and Chapters 1 and 3) and does not occur at an appreciable rate until far more negative potentials are reached. Thus, the first reduction of significance is the deposition of cadmium into the mercury to form the amalgam.

$$Cd^{2+} + 2e \rightleftharpoons Cd(Hg) \qquad \text{(actual first reduction, } \approx \text{-0.59 V vs. SCE)}$$

This is an example of kinetics superseding thermodynamic expectation during dynamic perturbation of an electrochemical system. The overpotential of hydrogen on mercury significantly widens the range of accessible potentials (i.e., the "potential window") in water, and is one reason mercury was long-favored for electrochemical analysis and thermodynamic evaluations despite its toxicity. The reduction waves for Cd^{2+} and Zn^{2+} can both be observed at a mercury electrode.

The reduction of Cd^{2+} will lead to a limiting current at -0.59 V vs SCE, followed by a limiting current of similar height for the reduction of Zn^{2+} at about -1.0 V vs. SCE. Finally, the reduction of H^+ begins at approximately -1.1 V vs. SCE with a cathodic current that appears infinite on the current scale due to the H^+ concentration of 1 M.

The current-potential curve resembles the following where the anodic current due to the oxidation of Hg at 0.0 V vs SCE appears practically infinite on the current scale. This oxidation defines the anodic background limit.

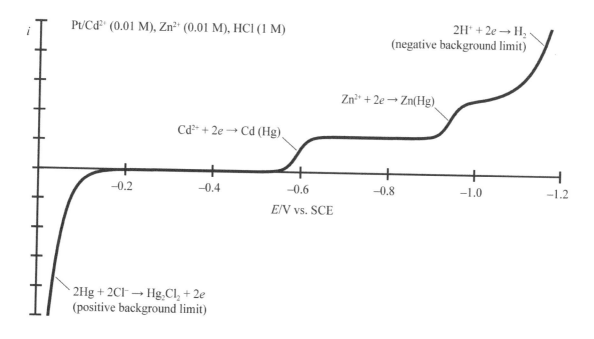

The negative background limit in this case is not due to a reversible or quasireversible couple, but, rather, the highly irreversible discharge of H_2 on a mercury electrode. It occurs at potentials about 1 V more extreme than E^0 for the couple because the kinetics have to be activated through a large overpotential.

Chapter 1   OVERVIEW OF ELECTRODE PROCESSES

Problem 1.3$^{\copyright}$ The important reactions are

$$Fe^{3+} + e \rightleftharpoons Fe^{2+} \qquad E^0 = 0.771 \text{ V vs. NHE}$$

$$Sn^{4+} + 2e \rightleftharpoons Sn^{2+} \qquad E^0 = 0.15 \text{ V vs. NHE}$$

(a). From (1.3.10) and $n = 1$, $i_l = nFAm_O C_O^* = 580 \ \mu A$.

(b).   Because the concentration of stannic ion is half that of ferric ion but $n = 2$, and the mass transfer coefficients of the two ions are the same, the limiting current for the reduction of Sn^{4+} is also 580 μA. The halfwave potential, $E_{1/2}$, for the ferric reduction is near $E^0 = 0.77$ V vs. NHE, whereas that for the reduction of stannic ion is near 0.15 V vs. NHE.

The $i - E$ curve is as follows.

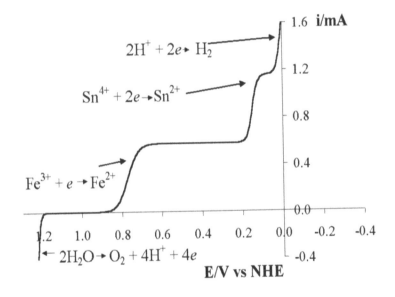

Problem 1.5[©]   The discussion for this problem is found in Section 1.6.4a. From equation (1.6.17),

$$i = -\frac{\Delta E}{R_u} \exp\left[-\frac{t}{R_u C_d A}\right] \tag{1.6.17}$$

Area A is introduced where C_d is expressed as capacitance per area (μF cm^{-2}). The cell time constant $\tau = R_u C_d A$.

$$i = -\frac{\Delta E}{R_u} \exp\left[-\frac{t}{\tau}\right]$$

For potential stepped from E_1 to E_2 in the nonfaradaic region, $\Delta E = E_2 - E_1$, the capacitor is charged. The initial current, when $t = 0$, is the maximum current, $= -\Delta E/R_u$. The current decays to $0.05 i\,(t = 0)$ at time $i\,(t_{5\,\%})$ when

$$\frac{i\,(t_{5\%})}{i\,(t = 0)} = 0.05 = \exp\left[-\frac{t_{5\%}}{\tau}\right]$$

$$-\ln(0.05) = 3.0 = \frac{t_{5\%}}{\tau}$$

The current decays to 5 % of the initial current when $t_{5\%} = 3.0\tau$. At this time, double-layer charging is 95 % complete.

For $C_d = 20\ \mu$F cm^{-2} and $A = 0.1$ cm^2, $\tau = R_u C_d A$ and $t_{5\%} = 3.0\tau$ are found for R_u of 1, 10, and 100 Ω.

R_s (Ω)	τ (μs)	3τ (μs)
1	2	6
10	20	60
100	200	600

=====

Problem 1.7[©]   From equation (1.6.23),

$$i = \nu C_d \left[1 - \exp\left(-\frac{t}{R_u C_d}\right)\right] \tag{1.6.23}$$

The transient term decays in a few time constants. Here, it is negligible after 10 μs (which corresponds to 5τ in Problem 1.5 at $R_s = 1\ \Omega$) to 2 ms (at $R_s = 100\ \Omega$). At steady state,

$$i = v C_d A$$

A appears as C_d is expressed here in capacitance per unit area. Currents are independent of R_u.

ν (V s^{-1})	0.02	1	20
i (μA)	0.04	2	40

Problem 1.8© **(a).** From equations (1.3.10) and (1.3.18) for the limiting currents, $\frac{i_{l,c}}{-i_{l,a}} = \frac{nFAm_OC_O^*}{nFAm_RC_R^*} = \frac{4.00\mu A}{2.40\mu A} = 1.67$ or $\frac{m_O}{m_R} = 1.67\frac{C_R^*}{C_O^*} = 0.833$. From equation (1.3.16), $E_{1/2} = E^{0'} - \frac{RT}{nF}\ln\frac{m_O}{m_R} = -0.498$ V vs. NHE.

(b). The wave has both cathodic and anodic parts, because both A^{3+} and A^+ are present in the bulk. The potential where the current is midway between the plateaus is $E_{1/2}$. The potential where $i = 0$ is the equilibrium value calculable from the Nernst equation. The $i - E$ curve is shown.

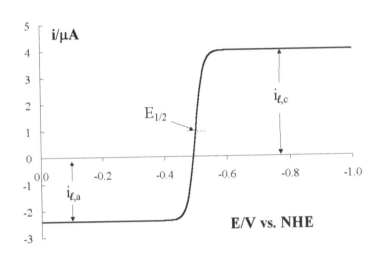

(c). From (1.3.21), the plot is $\log\left[(i_{l,c} - i)/(i - i_{l,a})\right]$ vs. NHE. The crossing of the abscissa is $E_{1/2}$, and the slope magnitude is $nF/(2.303RT) = n38.92$ V^{-1} at 25 °C. The plot is shown.

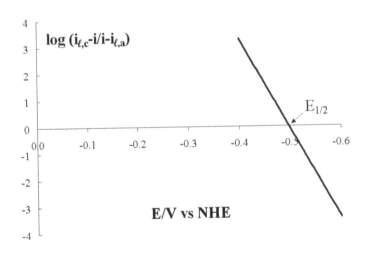

Problem 1.10© The relationships linking current and concentration in the steady-state treatment of mass transfer are equations (1.3.7) and (1.3.8).

$$i = nFAm_O[C_O^* - C_O(x=0)] \tag{1}$$

$$i = nFAm_R[C_R(x=0) - C_R^*] \tag{2}$$

Because $C_O^* = 0$, the first of these is

$$i = -nFAm_O C_O(x=0) \tag{3}$$

Because O does not exist in the bulk, no cathodic current can flow. All current goes to oxidize R. The limiting rate of oxidation is found when $C_R(x=0) = 0$, hence the limiting current is

$$i_{l,a} = -nFAm_R C_R^* \tag{4}$$

The system is reversible, hence,

$$E = E^{o\prime} + \frac{RT}{nF} \ln \frac{C_O(x=0)}{C_R(x=0)} \tag{5}$$

From equation (3),

$$C_O(x=0) = \frac{-i}{nFAm_O} \tag{6}$$

From equations (2) and (4),

$$C_R(x=0) = \frac{i - i_{l,a}}{nFAm_R} \tag{7}$$

Substitution of (6) and (7) into (5) gives

$$E = E^{0\prime} + \frac{RT}{nF} \ln \left[\frac{m_R}{m_O} \right] + \frac{RT}{nF} \ln \left[\frac{-i}{i - i_{l,a}} \right] \tag{8}$$

Note that this result is the special case of (1.3.21) for $i_{l,c} = 0$. When $i = i_{l,a}/2$, the last term in (8) is zero and $E = E_{1/2}$.

$$E_{1/2} = E^{0\prime} + \frac{RT}{nF} \ln \frac{m_R}{m_O} \tag{9}$$

The $i - E$ curve, plotted from equation (8), resembles the following:

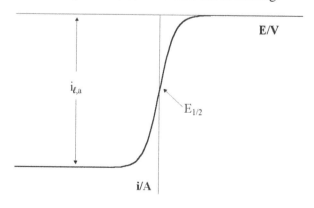

9

Chapter 1   OVERVIEW OF ELECTRODE PROCESSES

Problem 1.12©   **(a).** Starting with expression (1.3.17),

$$E = E_{1/2} + \frac{RT}{nF} \ln \left[\frac{i_l - i}{i} \right] \tag{1.3.17}$$

One solves for i/i_l as follows.

$$
\begin{aligned}
\frac{nF}{RT} \left(E - E_{1/2} \right) &= \ln \left[\frac{i_l - i}{i} \right] \\
\exp \left[\frac{nF}{RT} \left(E - E_{1/2} \right) \right] &= \frac{i_l - i}{i} = \frac{i_l}{i} - 1 \\
\frac{i_l}{i} &= 1 + \exp \left[\frac{nF}{RT} \left(E - E_{1/2} \right) \right] \\
\frac{i}{i_l} &= \left(1 + \exp \left[\frac{nF}{RT} \left(E - E_{1/2} \right) \right] \right)^{-1} \tag{1.11.1}
\end{aligned}
$$

(b). The limiting current, from expression (1.3.10), is

$$i_l = nFAm_O C_O^* \tag{1.3.10}$$

Here, $i_l = 96.5 \ \mu A$ for $A = 0.10 \ \text{cm}^2$, $m_O = 10^{-3} \ \text{cm/s}$, and $C_O^* = 10 \ \text{mM}$.

From Appendix C, $E^0 = 0.10 \ \text{V}$ vs. NHE. Note that because $m_O = m_R = 10^{-3} \ \text{cm/s}$, then $E_{1/2} = E^0$, which follows from equation (1.3.16). To calculate the $i - E$ curve, an Excel spreadsheet follows on page 11.

Calculation of Reversible Current-Potential Curves

Based on equation (1.3.17)

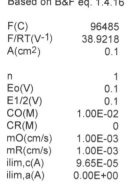

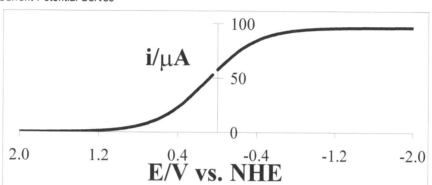

Calculation of Reversible Current-Potential Curves
Based on B&F eq. 1.4.16

F(C)	96485
F/RT(V^{-1})	38.9218
A(cm^2)	0.1
n	1
Eo(V)	0.1
E1/2(V)	0.1
CO(M)	1.00E-02
CR(M)	0
mO(cm/s)	1.00E-03
mR(cm/s)	1.00E-03
ilim,c(A)	9.65E-05
ilim,a(A)	0.00E+00

E/V	I/A	I/uA		E/V	I/A	I/uA
-2.00	9.646E-05	96.46		0.05	5.292E-05	52.92
-1.95	9.645E-05	96.45		0.10	4.824E-05	48.24
-1.90	9.644E-05	96.44		0.15	4.356E-05	43.56
-1.85	9.644E-05	96.44		0.20	3.897E-05	38.97
-1.80	9.643E-05	96.43		0.25	3.455E-05	34.55
-1.75	9.641E-05	96.41		0.30	3.036E-05	30.36
-1.70	9.640E-05	96.40		0.35	2.646E-05	26.46
-1.65	9.638E-05	96.38		0.40	2.289E-05	22.89
-1.60	9.636E-05	96.36		0.45	1.967E-05	19.67
-1.55	9.633E-05	96.33		0.50	1.680E-05	16.80
-1.50	9.629E-05	96.29		0.55	1.427E-05	14.27
-1.45	9.625E-05	96.25		0.60	1.206E-05	12.06
-1.40	9.620E-05	96.20		0.65	1.015E-05	10.15
-1.35	9.614E-05	96.14		0.70	8.514E-06	8.51
-1.30	9.607E-05	96.07		0.75	7.119E-06	7.12
-1.25	9.598E-05	95.98		0.80	5.938E-06	5.94
-1.20	9.588E-05	95.88		0.85	4.942E-06	4.94
-1.15	9.575E-05	95.75		0.90	4.105E-06	4.10
-1.10	9.559E-05	95.59		0.95	3.405E-06	3.40
-1.05	9.540E-05	95.40		1.00	2.820E-06	2.82
-1.00	9.517E-05	95.17		1.05	2.333E-06	2.33
-0.95	9.489E-05	94.89		1.10	1.929E-06	1.93
-0.90	9.456E-05	94.56		1.15	1.594E-06	1.59
-0.85	9.415E-05	94.15		1.20	1.316E-06	1.32
-0.80	9.366E-05	93.66		1.25	1.086E-06	1.09
-0.75	9.308E-05	93.08		1.30	8.953E-07	0.90
-0.70	9.238E-05	92.38		1.35	7.382E-07	0.74
-0.65	9.154E-05	91.54		1.40	6.085E-07	0.61
-0.60	9.055E-05	90.55		1.45	5.014E-07	0.50
-0.55	8.937E-05	89.37		1.50	4.131E-07	0.41
-0.50	8.797E-05	87.97		1.55	3.403E-07	0.34
-0.45	8.633E-05	86.33		1.60	2.803E-07	0.28
-0.40	8.443E-05	84.43		1.65	2.309E-07	0.23
-0.35	8.222E-05	82.22		1.70	1.901E-07	0.19
-0.30	7.969E-05	79.69		1.75	1.566E-07	0.16
-0.25	7.681E-05	76.81		1.80	1.289E-07	0.13
-0.20	7.359E-05	73.59		1.85	1.061E-07	0.11
-0.15	7.002E-05	70.02		1.90	8.738E-08	0.09
-0.10	6.613E-05	66.13		1.95	7.194E-08	0.07
-0.05	6.194E-05	61.94		2.00	5.923E-08	0.06
0.00	5.751E-05	57.51				

Chapter 1   OVERVIEW OF ELECTRODE PROCESSES

Problem 1.13[©]   **(a).** Combining equations (1.3.21) and (1.3.16) yields

$$E = E_{1/2} + \frac{RT}{nF} \ln \left[\frac{i_{l,c} - i}{i - i_{l,a}} \right]$$

This equation can be rewritten as

$$\frac{i_{l,c} - i}{i - i_{l,a}} = \exp \left[\frac{nF}{RT} \left(E - E_{1/2} \right) \right]$$

Or, upon solving for i,

$$i = \frac{i_{l,c} + i_{l,a} \exp \left[\frac{nF}{RT} \left(E - E_{1/2} \right) \right]}{1 + \exp \left[\frac{nF}{RT} \left(E - E_{1/2} \right) \right]}$$

(c). The equilibrium potential can be calculated from the Nernst equation, as follows.

$$E_{eq} = 0.10 \text{ V} + 0.05915 \, \log(10 \text{ mM}/5 \text{ mM}) = 0.12 \text{ V}$$

Substitution of $i = 0.48$ mA/cm^2 in equation (1.3.21) yields

$$E = 0.10 \text{ V} + 0.05915 \log \left[\frac{0.965 - 0.48}{0.48 - (-0.482)} \right] = 0.082 \text{ V}$$

From equation (1.2.1),

$$\begin{aligned} \eta_{mt} &= E - E_{eq} & (1.2.1) \\ \eta_{mt} &= 0.082 \text{ V} - 0.12 \text{ V} = -0.038 \text{ V} \end{aligned}$$

(d). The diffusion limiting cathodic current is 0.000965 A/cm$^2 \times 0.1$ cm$^2 = 9.65 \times 10^{-5}$ A. From equation (1.3.29), for a cathodic process,

$$R_{mt} = \frac{RT}{nFi_{l,c}} \qquad (1.3.29)$$

$R_{mt} = (RT/F) \left(ni_{l,c} \right)^{-1} = 0.0257 \text{ V} /9.65 \times 10^{-5} \text{ A} = 266 \, \Omega.$

2 POTENTIALS AND THERMODYNAMICS OF CELLS

Problem 2.1[©] According to the comments on page 4 of the text, the cell potential is a measure of the energy available to drive charge externally between the two electrodes; thus, the cell potential is positive. The standard emf is $E^0_{rxn} = E^0_{right} - E^0_{left}$, according to equation (2.1.42); under standard conditions, the emf yields the standard free energy as $\Delta G^0 = -nFE^0_{rxn}$, according to equation (2.2.26). If $\Delta G^0 < 0$, the reaction is spontaneous (galvanic); if $\Delta G^0 > 0$, the reaction must be driven by an external power source and the reaction is electrolytic.

(a). Any reaction pair with the form

Half Reactions:

$$O + \tfrac{q}{2}H_2O + ne \rightleftharpoons R + qOH^-$$

$$R + \tfrac{q}{2}H_2O \rightleftharpoons O + qH^+ + ne$$

has the net sum

Net Reactions:

$$H_2O \rightleftharpoons H^+ + OH^-$$

For example:

Half Reactions:

$$2H_2O + 2e \rightleftharpoons H_2 + 2OH^- \qquad E^0 = -0.828 \text{ V vs. NHE}$$

$$H_2 \rightleftharpoons 2H^+ + 2e \qquad E^0 = 0.000 \text{ V vs. NHE}$$

Net Reaction:

$$H_2O \rightleftharpoons H^+ + OH^-$$

Cell:

$$Pt / H_2(a = 1) / HCl(a = 1) // NaOH(a = 1) / H_2(a = 1) / Pt$$

Right electrode at $E^0_{H_2O/H_2} = -0.828$ V vs. NHE

Left electrode at $E^0_{H^+/H_2} = 0.000$ V vs. NHE

The right electrode is negative. The cell potential is 0.828 V. From equation (2.1.42), $E^0_{rxn} = E^0_{right} - E^0_{left} = -0.828$ V, so the cell must be operated electrolytically in carrying out the reaction.

(c).

Half Reactions:

$$PbSO_4 + 2e \rightleftharpoons Pb + SO_4^{2-} \qquad E^0 = -0.3505 \text{ V vs. NHE}$$

$$PbSO_4 + 2H_2O \rightleftharpoons PbO_2 + 4H^+ + SO_4^{2-} + 2e \qquad E^0 = 1.698 \text{ V vs. NHE}$$

Net Reactions:

$$2PbSO_4 + 2H_2O \rightleftharpoons PbO_2 + Pb + 4H^+ + 2SO_4^{2-}$$

Cell:

$$Pb / PbO_2 / PbSO_4 / H^+(a = 1), SO_4^{2-}(a = 1) / PbSO_4 / Pb$$

Right electrode at $E^0_{PbSO_4/Pb} = -0.3505$ V vs. NHE

Left electrode at $E^0_{PbO_2/PbSO_4} = 1.698$ V vs. NHE

Right electrode is negative. The cell potential is 2.048 V. From equation (2.1.42), $E^0_{rxn} = E^0_{right} - E^0_{left} = -2.048$ V, so the cell is electrolytic for the reaction as written. This is the charging reaction for a lead-acid battery.

Electrochemical Methods: Fundamentals and Applications, Third Edition, Student Solutions Manual. Cynthia G. Zoski and Johna Leddy.
© 2025 John Wiley & Sons Ltd. Published 2025 by John Wiley & Sons Ltd.

(d).

Half Reactions:	$TMPD^{\cdot} + e \rightleftharpoons TMPD$	$E^0 = 0.21$ V vs. NHE
	$An^{\cdot} \rightleftharpoons An + e$	$E^0 = -1.92$ V vs. NHE

Net Reactions: $TMPD^{\cdot} + An^{\cdot} \rightleftharpoons TMPD + An$

Cell: Pt / $An^{\cdot}(a = 1)$, $An(a = 1)$ in MeCN, 0.1 M TBAI //
$\qquad\qquad$ $TMPD^{\cdot}(a = 1)$, $TMPD(a = 1)$ in MeCN, 0.1 M TBAP / Pt

Right electrode at $E^0_{TMPD/TMPD^{\cdot}} = 0.21$ V vs. SCE

Left electrode at $E^0_{An/An^{\cdot}} = -1.92$ V vs. SCE

Left electrode is negative. The cell potential is 2.13 V. From equation (2.1.42), $E^0_{rxn} = E^0_{right} - E^0_{left} = 2.13$ V, so the cell is galvanic for the reaction as written. This ion annihilation is sufficiently energetic to produce chemiluminescence (Section 20.5).

(e).

Half Reactions:	$BQ + 2H^+ + 2e \rightleftharpoons H_2Q$	$E^0 = 0.6992$ V vs. NHE
	$2Ce^{3+} \rightleftharpoons 2Ce^{4+} + 2e$	$E^0 = 1.72$ V vs. NHE

Net Reactions: $2Ce^{3+} + BQ + 2H^+ \rightleftharpoons 2Ce^{4+} + H_2Q$

Cell: Pt / $Ce^{4+}(a = 1)$, $Ce^{3+}(a = 1)$ //
$\qquad\qquad$ $BQ(a = 1)$, $H_2Q(a = 1)$, $H^+(a = 1)$, $SO_4^{2-}(a = 1)$ / Pt

Right electrode at $E^0_{BQ,H_2Q} = 0.6992$ V vs. NHE

Left electrode at $E^0_{Ce^{4+},Ce^{3+}} = 1.72$ V vs. NHE

Right electrode is negative. The cell potential is 1.021 V. From equation (2.1.42), $E^0_{rxn} = E^0_{right} - E^0_{left} = -1.021$ V, so the cell is electrolytic for the reaction as written. Note that Ce^{4+} is among the most potent oxidants available in aqueous solutions.

=====

Problem 2.2[©] Standard potentials must be converted to free energies to calculate the correct standard potentials. Alternatively, tabulated values of the free energies can be used to calculate the standard potentials for the net half-cell reaction.

The approach is to convert the half reaction of interest into a full reaction through combination with an appropriate half reaction such as the hydrogen reduction half reaction.

$$H^+ + e \rightleftharpoons \frac{1}{2} H_2(g)$$

The free energy change is then calculated from the full reaction, which in turn yields E^0_{rxn} and the standard potential of the target half cell, E^0. The free energies needed to solve this problem are tabulated as follows.

Species	ΔG_f^0 (kcal/mol)	ΔG_f^0 (kJ/mol)
CO (g)	-32.81	-137.3
CO_2 (g)	-94.26	-394.6
CH_4 (g)	-12.14	-50.82
H_2O (l)	-56.69	-237.3
C_2H_2 (g)	50.00	209.3
C_2H_6 (g)	-7.86	-32.9
H_2 (g)	0.00	0.00

(a). The reaction

$$CO(g) + H_2O(l) \rightleftharpoons CO_2(g) + 2H^+ + 2e$$

is added to the hydrogen reduction half reaction to yield

$$CO(g) + H_2O(l) \rightleftharpoons CO_2(g) + H_2(g)$$

The standard free energy change for this reaction ΔG^0 is

$$\begin{aligned}
\Delta G^0 &= \Delta G_{f,CO_2}^0 + \Delta G_{f,H_2}^0 - \Delta G_{f,CO}^0 - \Delta G_{f,H_2O}^0 \\
&= -394.6 - 0.00 - (-137.3 - 237.3) = -20.0 \text{ kJ/mol}
\end{aligned}$$

Recall, the relationship between standard potential and free energy.

$$E_{rxn}^0 = -\frac{\Delta G^0}{nF} \tag{1}$$

For $n = 2$, $E_{rxn}^0 = -\frac{-20.0 \text{ kJ/mol}}{2 \times 96485 \text{ C/mol}} = 0.104$ V where $E_{rxn}^0 = E_{H^+/H_2}^0 - E_{CO_2/CO}^0$ or

$$E_{CO_2/CO}^0 = E_{H^+/H_2}^0 - E_{rxn}^o = 0.00 \text{ V} - 0.104 \text{ V} = -0.104 \text{ V vs. NHE}$$

Consider a cell where the cathode reaction is reduction of oxygen.

$$\tfrac{1}{2} O_2 + 2H^+ + 2e \rightleftharpoons H_2O \qquad 1.229 \text{ V vs. NHE}$$

(a').

$$\begin{array}{ll}
\tfrac{1}{2} O_2 + 2 H^+ + 2e \rightleftharpoons H_2O & 1.229 \text{ V} \\
-(CO(g) + H_2O(l) \rightleftharpoons CO_2(g) + 2H^+ + 2e) & -(-0.104 \text{ V}) \\
\hline
CO(g) + \tfrac{1}{2} O_2 \rightleftharpoons CO_2(g) + H_2O & 1.333 \text{ V} = E_{rxn}^0
\end{array}$$

To the extent that the cell can be discharged reversibly, E_{rxn}^0 would approach the cell voltage. The largest E_{rxn}^0 then corresponds to the largest cell voltage under standard conditions.

These ideas are important in the design of direct reformation fuel cells, where high current and high power are needed for a good power source. Thus, the thermodynamic choice of a direct reformation fuel would be dictated by the voltage and current (i.e., n), as well as by the power per weight and volume of the fuel. Note, the product of voltage and current sets the power output of the cell. (In direct reformation, fuel is fed directly to the cathode. Despite thermodynamic considerations, direct reformation fuel cells are not currently technologically viable because the cells are kinetically limited by passivation of the noble metal catalysts by adsorbed CO.) Issues of work available per weight are considered next.

At standard conditions, equation (1) is used to determine ΔG^0 from E^0_{rxn}, as tabulated.

Fuel	E^0_{rxn} (V)	n	ΔG^0 (kJ/mol)	ΔG^0 (kcal/mol)	MW	kJ/g	kcal/g
CO	1.332	2	-257.0	-61.4	28.01	-9.2	-2.2

$$=====$$

Problem 2.4$^©$ (a). Ag / AgCl / K^+, Cl^- (1M) / Hg_2Cl_2 / Hg

$$
\begin{array}{lll}
Hg_2Cl_2 + 2e \rightleftharpoons 2Hg + 2Cl^- & 0.26816 \text{ V} & = E^0_r \\
-2\times \left(AgCl + e \rightleftharpoons Ag + Cl^- \right) & -(0.2223 \text{ V}) & = E^0_l \\
\hline
Hg_2Cl_2 + 2Ag + 2Cl^- \rightleftharpoons 2Hg + 2Cl^- + 2AgCl & 0.0459 \text{ V} & = E^0_{rxn}
\end{array}
$$

$$Hg_2Cl_2 + 2Ag \rightleftharpoons 2Hg + 2AgCl \qquad \Delta G < 0; \text{ reaction is spontaneous}$$

(d). Pt / H_2(1 bar) / Na^+, OH^-(0.1 M) // Na^+, OH^- (0.1 M) / O_2 (0.2 bar) / Pt

$$
\begin{array}{lll}
O_2 + 2H_2O + 4e \rightleftharpoons 4OH^- & 0.401 \text{ V} & = E^0_r \\
-2\times(2H_2O + 2e \rightleftharpoons H_2 + 2OH^-) & -(0.828 \text{ V}) & = E^0_l \\
\hline
2H_2 + O_2 \rightleftharpoons 2H_2O & 1.229 \text{ V} & = E^0_{rxn}
\end{array}
$$

$$E_r = 0.401 + \frac{0.0591}{4} \log \frac{P_{O_2}}{[OH^-]^4} = 0.401 + 0.0148 \log \frac{[0.2]}{[0.1]^4} = 0.450 \text{ V}$$

$$E_l = -0.828 + \frac{0.0591}{2} \log \frac{1}{P_{H_2}[OH^-]^2} = -0.828 + 0.0296 \log \frac{1}{[1][0.1]^2} = -0.769 \text{ V}$$

$$E_{rxn} = E_r - E_l = 1.219 \text{ V (spontaneous)}$$

Alternatively, for the reaction as written,

$$E_{rxn} = 1.229 - \frac{0.0591}{4} \log \frac{1}{P^2_{H_2} P_{O_2}} = 1.229 - 0.0148 \log \frac{1}{[1]^2[0.2]} = 1.219 \text{ V}$$

Note that this cell reaction is the same as that in (c) and that the pressures of the gaseous reactants are also the same. Thus, E_{rxn} must be identical. However, the change in pH in the electrolyte does shift the potentials of the hydrogen and oxygen electrodes to more negative values by 59 mV per unit rise in pH. In practical terms, pH sets the accessible potentials or "solvent window" in aqueous solutions.

(f). Pt / Ce^{3+} (0.01 M), Ce^{4+} (0.1 M), H_2SO_4 (1 M) // Fe^{2+} (0.01 M), Fe^{3+} (0.1 M), HCl (1 M) / Pt

$$
\begin{array}{lll}
Fe^{3+} + e \rightleftharpoons Fe^{2+} & 0.771 \text{ V} & = E^0_r \\
-(Ce^{4+} + e \rightleftharpoons Ce^{3+}) & -(1.72 \text{ V}) & = E^0_l \\
\hline
Ce^{3+} + Fe^{3+} \rightleftharpoons Fe^{2+} + Ce^{4+} & -0.95 \text{ V} & = E^0_{rxn}
\end{array}
$$

$$E_r = 0.771 + 0.0591 \log \frac{[Fe^{3+}]}{[Fe^{2+}]} = 0.771 + 0.0591 \log \frac{[0.1]}{[0.01]} = 0.83 \text{ V}$$

$$E_l = 1.72 + 0.0591 \log \frac{[Ce^{4+}]}{[Ce^{3+}]} = 1.72 + 0.0591 \log \frac{[0.1]}{[0.01]} = 1.78 \text{ V}$$

$$E_{rxn} = E_r - E_l = -0.95 \text{ V (not spontaneous)}$$

Alternatively,

$$E_{rxn} = -0.95 - \log \frac{[Fe^{2+}][Ce^{4+}]}{[Ce^{3+}][Fe^{3+}]} = -0.95 - \log \frac{[0.01][0.1]}{[0.01][0.1]} = -0.95 \text{ V}$$

Problem 2.5[©] For the reaction,

$$\text{Ce}^{3+} + \text{Fe}^{3+} \rightleftharpoons \text{Ce}^{4+} + \text{Fe}^{2+}$$

From equation (2.1.42),

$$E^0_{rxn} = E^0_{Fe^{3+}/Fe^{2+}} - E^0_{Ce^{4+}/Ce^{3+}} = -0.95 \text{ V}$$

From equation (2.1.26),

$$\Delta G^0 = -nFE^0_{rxn} = 91.7 \text{ kJ/mole}$$

From equation (2.1.29) and at 25 °C,

$$K_{rxn} = K_{eq} = \exp\left[-\frac{\Delta G^0}{RT}\right] = 8.61 \times 10^{-17}$$

At all times, $[Ce^{3+}] + [Ce^{4+}] = 0.11$ M and $[Fe^{3+}] + [Fe^{2+}] = 0.11$ M.

Because E_{rxn} is negative [see Problem 2.4(f)], discharge converts Ce^{4+} and Fe^{2+} to Ce^{3+} and Fe^{3+}. Let n be the number of moles converted during discharge to equilibrium. Then, at equilibrium,

$[Ce^{3+}] = 0.01 + n/V_{left}$
$[Ce^{4+}] = 0.1 - n/V_{left}$
$[Fe^{2+}] = 0.01 - n/V_{right}$
$[Fe^{3+}] = 0.1 + n/V_{right}$

where V_{left} and V_{right} are the volumes of the two solutions respectively. For equal volumes, we can let $n/V = C$. With such a small equilibrium constant, one can expect the oxidation of Fe^{2+} by Ce^{4+} to go essentially to completion. Because Fe^{2+} is the limiting reagent, $C \approx 0.01$ M and

$[Ce^{3+}] = 0.02$ M
$[Ce^{4+}] = 0.09$ M
$[Fe^{3+}] = 0.11$ M
$[Fe^{2+}] = K_{eq}\frac{[Ce^{3+}][Fe^{3+}]}{[Ce^{4+}]} = 2.1 \times 10^{-18}$ M

The right electrode would be at
$\quad E_r = 0.771 + 0.0591\log\frac{[Fe^{3+}]}{[Fe^{2+}]} = 1.76$ V vs. NHE
Referring to Figure 2.1.2, $E = 1.52$ V vs. SCE.

The left electrode would be at
$\quad E_l = 1.72 + 0.0591\log\frac{[Ce^{4+}]}{[Ce^{3+}]} = 1.76$ V vs. NHE

Referring to Figure 2.1.2, $E = 1.52$ V vs. SCE.

Both electrodes are at the same potential, so the cell potential must be zero. Shorting the electrodes therefore produces no current flow, as must be true at equilibrium.

Chapter 2   POTENTIALS AND THERMODYNAMICS OF CELLS

Problem 2.6$^{©}$ The reaction of interest is $PbSO_4 \rightleftharpoons Pb^{2+} + SO_4^{2-}$.

$$
\begin{array}{lll}
PbSO_4 + 2e \rightleftharpoons Pb + SO_4^{2-} & -0.3505 \text{ V} & = E_r^0 \\
-(Pb^{2+} + 2e \rightleftharpoons Pb) & -(-0.1251 \text{ V}) & = E_l^0 \\
\hline
PbSO_4 \rightleftharpoons Pb^{2+} + SO_4^{2-} & -0.2254 \text{ V} & = E_{rxn}^0
\end{array}
$$

Cell: $Pb\,/\,Pb^{2+}(a=1),\ NO_3^-(a=1)\,//\,Na^+(a=1),\ SO_4^{2-}\,(a=1)\,/\,PbSO_4\,/\,Pb$

The cell reaction is the solubility equilibrium written above. From equations (2.1.42) and (2.1.29),

$$
\begin{aligned}
E_{rxn} &= E_{rxn}^0 = E_{PbSO_4/Pb}^0 - E_{Pb^{2+}/Pb}^0 = -0.3505 \text{ V} - (-0.1251 \text{ V}) = -0.2254 \text{ V} \\
\ln K_{sp} &= \frac{nFE_{rxn}^0}{RT} = -17.5 \text{ or } K_{sp} = 2.40 \times 10^{-8}
\end{aligned}
$$

=====

Problem 2.8$^{©}$ The cell reaction is

Half-Reactions: $AgCl + e(Cu') \rightleftharpoons Ag + Cl^-$

$Fe^{2+} \rightleftharpoons Fe^{3+} + e(Cu)$

Net   Reaction: $AgCl + Fe^{2+} + e(Cu') \rightleftharpoons Ag + Cl^- + Fe^{3+} + e(Cu)$

Because M is not involved in the overall reaction, it cannot affect the cell potential, which reflects ΔG^0 for this reaction. One can consider the cell at open circuit in terms of the species at equilibrium across various phase boundaries.

$$
\begin{aligned}
\bar{\mu}_e^{Cu} &= \bar{\mu}_e^M & (1) \\
\bar{\mu}_{Fe^{2+}}^S &= \bar{\mu}_{Fe^{3+}}^S + \bar{\mu}_e^M & (2) \\
\bar{\mu}_{Cl^-}^{AgCl} &= \bar{\mu}_{Cl^-}^S & (3) \\
\bar{\mu}_{Ag^+}^{AgCl} + \bar{\mu}_e^{Ag} &= \bar{\mu}_{Ag}^{Ag} & (4) \\
\bar{\mu}_e^{Ag} &= \bar{\mu}_e^{Cu'} & (5)
\end{aligned}
$$

Adding equations (2) to (4) gives

$$
\bar{\mu}_{Fe^{2+}}^S + \bar{\mu}_{Cl^-}^{AgCl} + \bar{\mu}_{Ag^+}^{AgCl} + \bar{\mu}_e^{Ag} = \bar{\mu}_{Fe^{3+}}^S + \bar{\mu}_{Cl^-}^S + \bar{\mu}_{Ag}^{Ag} + \bar{\mu}_e^M
$$

Substituting from (1) and (5) and recognizing that

$$
\bar{\mu}_{Ag^+}^{AgCl} + \bar{\mu}_{Cl^-}^{AgCl} = \bar{\mu}_{AgCl}^{AgCl}
$$

gives

$$
\bar{\mu}_{Fe^{2+}}^S + \bar{\mu}_{AgCl}^{AgCl} + \bar{\mu}_e^{Cu'} = \bar{\mu}_{Fe^{3+}}^S + \bar{\mu}_{Ag}^{Ag} + \bar{\mu}_{Cl^-}^S + \bar{\mu}_e^{Cu}
$$

18

Expansion gives

$$\mu_{Fe^{2+}}^{0S} + RT \ln a_{Fe^{2+}}^{S} + 2F\phi^{S} + \mu_{AgCl}^{0AgCl} + \mu_{e}^{0Cu'} - F\phi^{Cu'}$$
$$= \mu_{Fe^{3+}}^{0S} + RT \ln a_{Fe^{3+}}^{S} + 3F\phi^{S} + \mu_{Ag}^{0Ag} + \mu_{Cl^-}^{0S} + RT \ln a_{Cl^-}^{S} - F\phi^{S} + \mu_{e}^{0Cu} - F\phi^{Cu}$$

Rearrangement provides

$$\phi^{Cu'} - \phi^{Cu} = E = \frac{\mu_{Fe^{2+}}^{0S} + \mu_{AgCl}^{0AgCl} - \mu_{Fe^{3+}}^{0S} - \mu_{Ag}^{0Ag} - \mu_{Cl^-}^{0S}}{F} + \frac{RT}{F} \ln \frac{a_{Fe^{2+}}^{S}}{a_{Fe^{3+}}^{S} a_{Cl^-}^{S}}$$

No terms describing energies in M appear, thus E does not depend on M. The term $\bar{\mu}_{e}^{M}$ appeared in equations (1) and (2) above, but they cancelled out. Distinct values of the interfacial potential difference, $\phi^{Cu} - \phi^{M}$, would arise for various species M, but the variations would be exactly compensated by variations in $\phi^{M} - \phi^{S}$.

$$=====$$

Problem 2.10 Consistent with the comments at the start of Problem 2.2, a sound thermodynamic development of standard potentials (E^0) for half-cell reactions must proceed through free energy calculations, not standard potentials.

(a). First, convert the two standard half-cell reactions involving Cu into net reactions by combining with the H^+/H_2 half-cell reaction.

$$H^+ + e \rightleftharpoons \tfrac{1}{2} H_2(g)$$

Thus,

$$
\begin{array}{lll}
Cu^{2+} + 2e \rightleftharpoons Cu & 0.340 \text{ V} & = E_r^0 \\
-2(H^+ + e \rightleftharpoons \tfrac{1}{2} H_2(g)) & -(0.000 \text{ V}) & = E_l^0 \\
\hline
Cu^{2+} + H_2 \rightleftharpoons Cu + 2H^+ & 0.340 \text{ V} & = E_{rxn,1}^0
\end{array}
$$

$$\Delta G_1^0 = -nFE_{rxn,1}^0 = -2FE_{rxn,1}^0$$

and

$$
\begin{array}{lll}
Cu^{2+} + I^- + e \rightleftharpoons CuI & 0.86 \text{ V} & = E_r^0 \\
-(H^+ + e \rightleftharpoons \tfrac{1}{2} H_2(g)) & -(0.00 \text{ V}) & = E_l^0 \\
\hline
Cu^{2+} + I^- + \tfrac{1}{2} H_2 \rightleftharpoons CuI + H^+ & 0.86 \text{ V} & = E_{rxn,2}^0
\end{array}
$$

$$\Delta G_2^0 = -nFE_{rxn,2}^0 = -1FE_{rxn,2}^0$$

Then, note that subtracting the second reaction from the first yields

$$CuI + \tfrac{1}{2} H_2 \rightleftharpoons Cu + I^- + H^+$$

This has a standard free energy of

$$\Delta G^0 = \Delta G_1^0 - \Delta G_2^0 = -F \left(2E_{rxn,1}^0 - E_{rxn,2}^0 \right)$$

This is a single electron transfer reaction, so the emf for this reaction is $E_{rxn,3}^0 = -\frac{\Delta G^0}{1F} = 2E_{rxn,1}^0 - E_{rxn,2}^0 = -0.18$ V.

Finally, the standard potential for the target half reaction is found as

$$
\begin{array}{ll}
CuI + e \rightleftharpoons Cu + I^- & = E_r^0 \\
-(H^+ + e \rightleftharpoons \frac{1}{2} H_2(g)) & -(0.00\ V) = E_l^0 \\
\hline
CuI + \frac{1}{2} H_2 \rightleftharpoons Cu + I^- + H^+ & -0.18\ V = E_{rxn,3}^0
\end{array}
$$

which is satisfied for the standard potential of the half reaction, $E_r^0 = E_{rxn,3}^0 + E_l^0 = -0.18$ V.

Generalized Form: The above processes can be generalized and simplified because the reference half reaction of H^+/H_2 and $-F$ cancel out. For the addition or subtraction of the standard potentials (E_1^0 and E_2^0) of two half reactions to yield the standard potential (E_3^0) of a third half reaction,

$$
E_3^0 = \frac{n_1 E_1^0 \pm n_2 E_2^0}{n_3} \tag{1}
$$

where the reactions have n_1, n_2, and n_3 electrons, respectively. Note that in all the previous problems in this Chapter, the special case applies where reactions are combined to yield a net equation with no explicit electrons. Then, $n_1 = n_2 = n_3$, and equation (1) reduces to $E_3^0 = E_1^0 \pm E_2^0$.

(b) This example is done using the generalized expression, equation (1). The half reactions are combined by subtraction of Rxn2 from Rxn1. Note, that the calculations yield a half reaction (i.e., there are explicit electrons in the final reaction) and the generalized form is required.

$$
\begin{array}{ll}
O_2 + 4H^+ + 4e \rightleftharpoons 2H_2O & E_1^0 = 1.229\ V \\
-(H_2O_2 + 2H^+ + 2e \rightleftharpoons 2H_2O) & E_2^0 = -(1.763\ V) \\
\hline
O_2 + 2H^+ + 2e \rightleftharpoons H_2O_2 & E_3^0
\end{array}
$$

where

$$
E_3^0 = \frac{4E_1^0 - 2E_2^0}{2} = \frac{4 \times 1.229 - 2 \times 1.763}{2} = 0.695\ V
$$

=====

Problem 2.12$^{\copyright}$ The total charge passed through the cell consists of the two components representing ionic (q_{ion}) and electronic (q_{el}) conduction.

$$
q = q_{ion} + q_{el}
$$

The q_{ion} component is due to a faradaic process (i.e., reduction of silver) and can be calculated as follows for $q_{ion} = nF \times moles$ where here $n = 1$.

$$
q_{ion} = 1 \times 96485\ C/mol \times \frac{1.12\ g - 1.00\ g}{107.87\ g/mol} = 107\ C
$$

The total charge passed is $q = 0.2A \times 600\ s = 120\ C$. Thus,

$$
q_{el} = 120\ C - 107\ C = 13\ C
$$

and

$$
\frac{q_{el}}{q} = \frac{13\ C}{120\ C} = 0.11
$$

gives the fraction of the current passing through the cell due to electronic conduction.

Problem 2.14© **(a).** Type 2, common anion. From equation (2.3.39),

$$E_j = -\frac{RT}{F} \ln \frac{\Lambda_{NaCl}}{\Lambda_{HCl}}$$

From equation (2.3.14) and using Table 2.3.2,

$$\begin{aligned}
\Lambda_{NaCl} &= \lambda_{Na^+} + \lambda_{Cl^-} = 50.11 + 76.34 = 126.45 \\
\Lambda_{HCl} &= \lambda_{H^+} + \lambda_{Cl^-} = 349.82 + 76.34 = 426.16
\end{aligned}$$

Substitution leads to $E_j = 31.2$ mV. The junction is dominated by the very mobile H^+, which tends to place a positive net charge in the right hand phase.

(c). Type 3. From the Hendersen equation (2.3.38), $E_j = 46.2$ mV. The junction is dominated by mobile OH^- which deposits a net negative charge on the left- hand phase. The situation is analogous to (b), but OH^- is not as mobile as H^+, hence E_j is lower here than in (b).

=====

Problem 2.16© By analogy to equation (2.4.20), the cell potential can be expressed in terms of concentrations, C_j, when activity coefficients are unity and the standard state concentration is 1 M:

$$E = constant + \frac{RT}{F} \ln \left(C_{Na^+} + k_{Na^+,j}^{pot} C_j \right)$$

where j is the interferent. The cell potential is always the value expected for $C_{Na^+} = 10^{-3}$ M in the absence of interference. For a 10% error, $k_{Na^+,j}^{pot} C_j$ must be 1×10^{-4} M.

The interferent concentrations that would cause this error are

$$\begin{aligned}
K^+: \quad & C_{K^+} = \frac{1\times 10^{-4}}{k_{Na^+,K^+}^{pot}} = \frac{1\times 10^{-4}}{0.001} = 0.1 \text{ M} \\
NH_4^+: \quad & C_{NH_4^+} = \frac{1\times 10^{-4}}{k_{Na^+,NH_4^+}^{pot}} = \frac{1\times 10^{-4}}{10^{-5}} = 10 \text{ M} \\
Ag^+: \quad & C_{Ag^+} = \frac{1\times 10^{-4}}{k_{Na^+,Ag^+}^{pot}} = \frac{1\times 10^{-4}}{300} = 3.3 \times 10^{-7} \text{ M} \\
H^+: \quad & C_{H^+} = \frac{1\times 10^{-4}}{k_{Na^+,H^+}^{pot}} = \frac{1\times 10^{-4}}{100} = 1 \times 10^{-6} \text{ M}
\end{aligned}$$

Problem 2.20[©]   (a). From equation (2.1.49)

$$
\begin{aligned}
E^{0\prime} &= E^0 + \frac{RT}{nF} \ln \frac{\gamma_{Fe^{3+}}}{\gamma_{Fe^{2+}}} \\
&= 0.771 \text{ V} + 0.02569 \text{ V} \ln \frac{0.179}{0.401} \\
&= 0.771 \text{ V} + 0.02569 \text{ V} \, (-0.807) \\
&= 0.750 \text{ V vs NHE in 0.1 M HClO}_4
\end{aligned}
$$

(b). From Table C.2, $E^{0\prime} = 0.735$ V vs NHE in 1 M HClO$_4$. From equation (2.1.49):

$$
\begin{aligned}
\ln \frac{\gamma_{Fe^{3+}}}{\gamma_{Fe^{2+}}} &= \frac{nF}{RT} \left(E^{0\prime} - E^0 \right) \\
\exp \left[\ln \frac{\gamma_{Fe^{3+}}}{\gamma_{Fe^{2+}}} \right] &= \exp \left[\frac{nF}{RT} \left(E^{0\prime} - E^0 \right) \right] \\
\frac{\gamma_{Fe^{3+}}}{\gamma_{Fe^{2+}}} &= \exp \left[38.92 \text{ V}^{-1} \, (0.735 - 0.771) \text{ V} \right] \\
&= 0.246 \text{ in 1 M HClO}_4
\end{aligned}
$$

This compares to part (a), where $\frac{\gamma_{Fe^{3+}}}{\gamma_{Fe^{2+}}} = \frac{0.179}{0.401} = 0.446$ in 0.1 M HClO$_4$.

At 1 M, the activity coefficient for Fe^{3+} is even smaller relative to that for Fe^{2+}. The trends in Table 2.1.1 found below 0.1 M (i.e., $\gamma_{Fe^{3+}}/\gamma_{Fe^{2+}}$ is 0.655 at 0.01 M and 0.847 at 0.01 M) increase monotonically up to 1 M. The higher charge of Fe^{3+} contracts the ionic atmosphere around the trication to lower the free energy of Fe^{3+} more than that of Fe^{2+}. The higher the ionic strength, the greater the suppression of $\gamma_{Fe^{3+}}/\gamma_{Fe^{2+}}$.

(c). From Table C.2, $E^{0\prime} = 0.70$ V vs NHE in 1 M HCl, but $E^{0\prime} = 0.53$ V vs NHE in 10 M HCl.

From Table C.1,

$$
Fe^{3+} + e \rightleftharpoons Fe^{2+} \qquad E^0 = 0.771 \text{ V vs NHE}
$$

Compared to the standard potential E^0, the shift in formal potential $E^{0\prime}$ in 1 M HCl is due to the complexation of the iron cations by Cl$^-$ with stronger complexation of Fe^{3+} than Fe^{2+}.

3 BASIC KINETICS OF ELECTRODE REACTIONS

Problem 3.1[©] **(a).** From equation (3.4.6),

$$j_0 = Fk^0 C_O^{*(1-\alpha)} C_R^{*\alpha} \tag{3.4.6}$$

$$= \left(96485 \frac{C}{mol}\right) \left(10^{-7} \frac{cm}{s}\right) \left(1 \times 10^{-6} \frac{mol}{cm^3}\right)^{(1-0.3)} \left(1 \times 10^{-6} \frac{mol}{cm^3}\right)^{0.3}$$

$$= 1 \times 10^{-8} \text{ A/cm}^2 = 0.01 \ \mu\text{A/cm}^2$$

(b) and (c). Because mass transfer effects are neglected, from equation (3.4.11),

$$j = j_0(\exp[-\alpha f\eta] - \exp[(1-\alpha)f\eta]) \tag{3.4.11}$$

where $f = F/RT$. An Excel spreadsheet can be set up to calculate j for a given η value. Such a spreadsheet is shown below. The current density-overpotential curve for this reaction and the corresponding Tafel plot are also shown. Note that for the Tafel plot, the curves are linear for $|\eta| > 100$ mV. For less extreme overpotentials, both the anodic and cathodic terms contribute to the current and the linearization is lost.

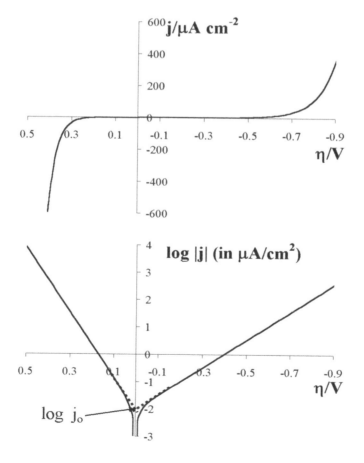

Chapter 3   BASIC KINETICS OF ELECTRODE REACTIONS

Problem 3.3[©]   **(a).** Current is expressed in terms of overpotential η and the exchange current density i_0 in equation (3.4.30).

$$i = \frac{\exp\left[-\alpha f\eta\right] - \exp\left[(1-\alpha)\,f\eta\right]}{\frac{1}{i_0} + \frac{\exp[-\alpha f\eta]}{i_{l,c}} - \frac{\exp[(1-\alpha)f\eta]}{i_{l,a}}} \tag{3.4.30}$$

A spreadsheet is shown for input potential E (V) that is used to calculate $\eta = E - E^{0\prime}$ where standard potential $E^{0\prime} = -0.5$ V vs NHE. Exchange current is defined in equation (3.4.6), as $i_0 = FAk^0 C_O^{*(1-\alpha)} C_R^{*\,\alpha}$. Parameters specified here are $A = 1.$ cm^2; $k^0 = 1.0 \times 10^{-4}$ cm/s; $\alpha = 0.5$; $C_O^* = 1.0 \times 10^{-3}$ mol/cm^3, and $C_R^* = 1.0 \times 10^{-5}$ mol/cm^3. $n = 1$. Mass transport parameters are given as $m_O = m_R = 0.01$ cm/s. The limiting currents are defined in equation (1.3.10) as $i_{l,c} = nFAm_O C_O^*$ and in equation (1.3.18) as $i_{l,a} = -nFAm_R C_R^*$. Current i is calculated on substitution of i_0, $i_{l,c}$, $i_{l,a}$, α, and η into equation (3.4.30). For these conditions, $i_0 = 9.6 \times 10^{-4}$ A; $i_{l,c} = 0.96$ A; and $i_{l,a} = 9.6 \times 10^{-3}$ A.

Equation (3.4.30) includes mass transport effects. The spreadsheet and figures are shown. The current voltage plot exhibits limiting currents set by mass transport. The plot of $\ln |i|$ versus η yields i_0 at $\eta = 0$; at more extreme η, $\ln |i|$ is constant due to mass transport limitations. All values of E are relative to NHE.

E/V	η/V	i/A	$\ln i$ (in A)
-1.50	-1.00	0.96485	-0.0358
-1.45	-0.95	0.96484	-0.0358
-1.40	-0.90	0.96483	-0.0358
-1.35	-0.85	0.96479	-0.0358
-1.30	-0.80	0.96468	-0.0360
-1.25	-0.75	0.96441	-0.362
-1.20	-0.70	0.96369	-0.0370
-1.15	-0.65	0.96178	-0.0390
-1.10	-0.60	0.95677	-0.0442
-1.05	-0.55	0.94375	-0.0579
-1.00	-0.50	0.91094	-0.0933
-0.95	-0.45	0.83417	-0.1813
-0.90	-0.40	0.68202	-0.3827
-0.85	-0.35	0.45992	-0.7767
-0.80	-0.30	0.24697	-1.3985
-0.75	-0.25	0.11091	-2.1990
-0.70	-0.20	0.04506	-3.0998
-0.65	-0.15	0.01743	-4.0496
-0.60	-0.10	0.00649	-5.0382
-0.55	-0.05	0.00210	-6.1636
-0.50	0.00	0.00	$-\infty$
-0.45	0.05	-0.00173	-6.3592
-0.40	0.10	-0.00389	-5.5483
-0.35	0.15	-0.00625	-5.0751
-0.30	0.20	-0.00801	-4.8267
-0.25	0.25	-0.00896	-4.7151
-0.20	0.30	-0.00938	-4.6696
-0.15	0.35	-0.00954	-4.6519
-0.10	0.40	-0.00961	-4.6451
-0.05	0.45	-0.00963	-4.6425
0.00	0.50	-0.00964	-4.6415

24

(b). k^0 values ranging from 1.0×10^{-4} to 10 cm/s were used. $\alpha = 0.5$. From the graph, one can see that for k^0 values > 0.1 cm/s under the conditions of part (a), the curves are indistinguishable from nernstian behavior.

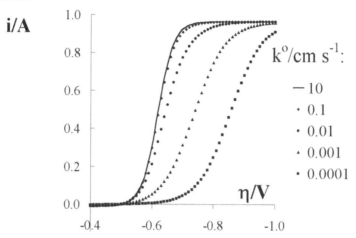

(c). The current-overpotential plot steepens as α increases. For $O + e \rightleftharpoons R$, transfer coefficient α describes the partition of energy $F\eta$ into the cathodic process ($O + e \longrightarrow R$) at the transition state, as in equation (3.3.4), $\Delta G_c^{\ddagger} = \Delta G_{0c}^{\ddagger} + \alpha F \left(E - E^{0\prime} \right)$. (Implicit in R. Parsons, Croatica Chimica ACTA 42 (1970) 281-290, $\alpha = \partial \Delta G^{\ddagger}/\partial \eta$) As α increases, the energy partitioned to the reduction increases and so the rate expressed as current increases with η.

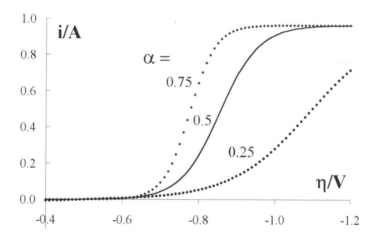

Chapter 3   BASIC KINETICS OF ELECTRODE REACTIONS

Problem 3.4©   (a). With no mass transfer effects, the current is given by the Butler-Volmer equation in equation (3.4.11).

$$i_{BV} = i_0 \left[\exp\left(-\alpha f \eta\right) - \exp\left((1 - \alpha) f \eta\right) \right] \tag{3.4.11}$$

The series expansion $e^x = \sum_{j=0}^{\infty} \frac{x^j}{j!} = 1 + x + \frac{x^2}{2!} + \frac{x^3}{3!} + \dots$. For $|x|$ small, $e^{-x} \approx 1 - x$ and $e^x \approx 1 + x$. Then, $i \to i_0 \left[1 - \alpha f \eta - (1 + (1 - \alpha) f \eta) \right]$. This derives equation (3.4.12) for sufficiently small η.

$$i = -i_0 f \eta \quad \text{for sufficiently small } \eta \tag{3.4.12}$$

Let the approximation for i in equation (3.4.12) be denoted i_{approx}.

The relative error ε of the linear characteristic relative to the Butler Volmer equation is found.

$$\varepsilon = \frac{i_{approx} - i_{BV}}{i_{BV}} = \frac{i_{approx}}{i_{BV}} - 1 = \frac{-f\eta}{\exp\left[-\alpha f \eta\right] - \exp\left[(1 - \alpha) f \eta\right]} - 1$$

Relative errors are tabulated below. The magnitude of the relative error increases as $|\eta|$ increases, which is consistent with the constraint that $|\eta|$ is sufficiently small that $e^{\pm x}$ is well approximated by $1 \pm x$. Tabulated relative errors are larger for $\alpha = 0.10$ than for $\alpha = 0.5$. Errors are symmetric about $\eta = 0$ for $\alpha = 0.5$, but asymmetric for $\alpha = 0.1$. The approximation that η is sufficiently small is more readily satisfied for the oxidation and the reduction where $\alpha \approx 1 - \alpha$. Equation (3.4.11) is equivalently expressed in equation (3.4.18) as $i_{BV}(\eta)/i_0 = \exp\left(-\alpha f \eta\right)\left[1 - \exp\left(f\eta\right)\right]$. If $\alpha \approx 0.5$, $i_{BV}(\eta)/i_0$ responds symmetrically to polarization of $\eta > 0$ and $\eta < 0$. But for $\alpha = 0.1$ and the same magnitude ($|\eta|$) overpotential, $i_{BV}(\eta)/i_0$ is more sensitive to $\eta < 0$ than $\eta > 0$. This is consistent with the asymmetry of the transition state energies, as reflected in $\alpha = 0.1$.

Relative Error for Linear $i - \eta$ Characteristic

η, mV	ε, % for $\alpha = 0.50$	ε, % for $\alpha = 0.10$
-50	-14.2	86.9
-20	-2.5	33.1
-10	-0.63	16.1
10	-0.63	-15.0
20	-2.5	-28.6
50	-14.2	-60.6

Problem 3.6[©]   **(a).** The rate expression for the first order rate reaction is

$$\frac{dn_A(t)}{dt} = -k_f n_A(t) \tag{1}$$

For an initial concentration of $n_{A,0}$, this is solved by separation of variables as

$$-k_f \int_0^t dt = \int_{n_{A,0}}^{n_A(t)} \frac{1}{n_A(t)} dn_A(t) = \ln n_A(t)|_{n_{A,0}}^{n_A(t)} = \ln \frac{n_A(t)}{n_{A,0}} = -k_f t$$

or

$$n_A(t) = n_{A,0} \exp\left[-k_f t\right] \tag{2}$$

At time t, some fraction of the molecules dn_A decay. By averaging the times over all molecules and normalizing by $n_{A,0}$, the average lifetime is determined.

$$\tau = \frac{\int_0^{n_{A,0}} t\, dn_A(t)}{n_{A,0}}$$

From (1) and (2) above,

$$dn_A(t) = -k_f n_A(t) dt = -k_f n_{A,0} \exp\left[-k_f t\right] dt$$

Then,

$$\tau = \frac{\int_0^\infty -t k_f n_{A,0} \exp\left[-k_f t\right] dt}{n_{A,0}} = -k_f \int_0^\infty t \exp\left[-k_f t\right] dt$$

This is integrated by parts where $\int u\, dv = uv - \int v\, du$. Let $u = t$ and $dv = \exp\left[-k_f t\right] dt$, such that $du = dt$ and $v = \frac{-\exp\left[-k_f t\right]}{k_f}$.

$$\begin{aligned}
\tau &= -k_f \int_0^\infty t \exp\left[-k_f t\right] dt = -k_f \left(\frac{-t \exp\left[-k_f t\right]}{k_f}\Big|_0^\infty - \int_0^\infty \frac{-\exp\left[-k_f t\right]}{k_f} dt \right) \\
&= -\int_0^\infty \exp\left[-k_f t\right] dt = -\frac{\exp\left[-k_f t\right]}{k_f}\Big|_0^\infty = \frac{1}{k_f}
\end{aligned}$$

Chapter 3   BASIC KINETICS OF ELECTRODE REACTIONS

(b). Consider a zone of thickness d from the electrode surface in which all molecules of species O are confined and can be reduced by electron transfer from the electrode. Let the electrode area be A. Then, the number of moles remaining at time t, $N_O(t)$, is

$$N_O(t) = AdC_O(0,t)$$

$C_O(0,t)$ is the confined concentration of the oxidized species at time t.

The decay rate is

$$\frac{dN_O(t)}{dt} = -kN_O(t) \qquad (moles/s)$$

where k is a first order decay constant (s^{-1}) applicable to the whole confined population of species O.

Consider the reaction rate per unit area, v_f.

$$v_f = \frac{1}{A}\frac{dN_O(t)}{dt} = -\frac{kN_O(t)}{A} = kdC_O(0,t) \qquad (moles/(cm^2 s))$$

The usual expression for the heterogeneous kinetics of the reaction O + e $\rightleftharpoons$ R *is*

$$v_f = k_f C_O(0,t) \qquad (3.2.5)$$

Because the reduction of O can happen only at the surface, one can equate these two rate expressions. Thus, $k_f = k_d$. The lifetime of the surface-confined system is $\tau = 1/k = d/k_f$, which should also be an approximation of the mean lifetime of species O arriving at the surface in a normal diffusional system. However, the lifetime of a given O arriving by diffusion to the electrode will likely vary. There is a range of electron transfer rates for the reduction of O at the surface because the rate of electron transfer varies with distance from the electrode surface. (See Section 3.5.2).

(c). For a lifetime of 1 ms, given $d = 10$ Å $= 1$ nm, $k_f = 10^{-7}$ cm/10^{-3} s $= 10^{-4}$ cm/s. For $\tau = 1$ ns, $k_f = 100$ cm/s. Such rates are readily accessible in electrochemical systems. For cyclic voltammetric perturbations at macroscopic electrodes and scan rates of about 100 mV/s, standard $(E = E^{0\prime})$ heterogeneous electron transfer rates > 0.1 cm/s are considered fast (reversible) and rates $< 10^{-5}$ cm/s are considered very slow (irreversible). See Chapter 7. With microelectrodes, standard heterogeneous rates in the range of 10 to 40 cm/s have been measured (see page 131). Note, however, that k_f is k^0 amplified by electrode overpotential through $\exp\left[-\alpha f\left(E - E^{0\prime}\right)\right]$, a term which under typical room temperature conditions exceeds 100 for an overpotential of 250 mV and 16000 for 500 mV. Thus, lifetimes as short as 1 ns should be possible, favored by high standard heterogeneous rates and large overpotentials.

Problem 3.7© When an electrode is immersed in a solution, it generally enters at some potential other than the equilibrium value. (In fact, it is usually regarded as entering at the PZC). Suppose this value is more positive than the equilibrium value dictated by the Nernst equation from the concentrations of Fe(II) and Fe(III) present in solution. The species Fe(II) and Fe(III) respond to the potential by undergoing charge transfer with the electrode at rates manifested by the component currents i_a and i_c. If the potential is more positive than E_{eq}, then oxidation of Fe(II) is favored over reduction of Fe(III), and i_a exceeds i_c. There is a net flow of electrons from solution to electrode. If the electrons are not drained away through an external circuit, then the electrons are left on the electrode. The charge on the electrode side of the double layer therefore becomes more negative, and the potential moves correspondingly negative. As E moves towards E_{eq} from the positive side, the imbalance between i_a and i_c lessens, hence the rate of change of q^M lessens, and the potential also changes more slowly. However, it continues to shift negatively as long as E exceeds E_{eq}. When E reaches E_{eq}, $i_a = i_c$ and the charge on the electrode changes no further. An exactly analogous scenario can be set forth for an initial potential more negative than E_{eq}, but in that case $i_c > i_a$ and the values of q^M and E shift positively.

For a typical capacitance of 20 $\mu F/cm^2$, a 1 V shift in potential requires 20 $\mu C/cm^2$. Thus, a 100 mV shift requires a change of 2 $\mu C/cm^2$ in q^M. Note that this charge represents a *net faradaic reaction* in the process of coming to equilibrium. In the iron system under consideration here, with $A = 0.1$ cm^2.

$$\text{Moles of Fe(II) to be oxidized} = \frac{2 \times 10^{-6} \text{ C/cm}^2 \times 0.10 \text{ cm}^2}{96485 \text{ C/mol}} = 2 \times 10^{-12} \text{ moles}$$

Thus, $\approx 10^{-12}$ moles of Fe(II) would have to be oxidized to shift the potential of a 0.10 cm^2 electrode to equilibrium from a potential 100 mV more positive.

At very low concentrations, e.g. 10^{-10} M or 10^{-11} M , only 10^{-11} moles of Fe(II) might exist in the whole of a typical test volume of 100 mL. In this case the iron species in solution lack the *capacity* to enforce an equilibrium on an electrode of 0.10 cm^2 area, where electrolyte, $\lesssim 10^{-11}$ M is insufficient. Loss of sensitivity occurs at considerably higher concentrations, however. Even at 10^{-7} to 10^{-8} M, transport of species to the electrode may not suffice to achieve equilibrium in a reasonable time. Moreover, the rate of potential change becomes slow as the charge-delivering capacity of the solution falls with concentration, and it becomes harder to tell whether equilibrium has been reached. Finally, it must be realized that other redox species exist in the system, e.g., O_2, H_2O, Cl^-, Pt, PtO. Other redox species also respond to the potential and deliver small currents to the interface. In fact, these currents may exceed those of Fe(II) and Fe(III) when the concentrations of the iron species become too low. The electrode will drift to the null current potential, where the i_a's from all sources cancel the i_c's from all sources. This potential is called a mixed potential because several couples may contribute to its definition (see Section 3.6). These considerations all contribute to the experimental fact that potentiometric measurements become unreliable at analyte concentrations below about 10^{-5} M.

The loss of sensitivity is a *kinetic problem*. The lack of a thermodynamic response comes about because there is no valid kinetic mechanism to establish thermodynamic equilibrium.

These same concepts apply to ion-selective electrodes (ISE). Registration of a potential requires a change in the state of charge of an interface. In the case of an ISE, the change in charge state usually comes about by selective adsorption or desorption. At sufficiently low analyte concentrations, the system becomes incapable of delivering the necessary charge, or competing, less selective, adsorption processes become important.

Problem 3.8©   The tabulated data show that a limiting cathodic current $i_{l,c} = 965$ μA is reached at $\eta = -500$ and $\eta = -600$ mV. Comparatively large overpotentials are required to enforce this current; hence, Tafel behavior should be observed for currents less than $\approx 10\%$ of the limiting current. A Tafel plot of the data is shown below.

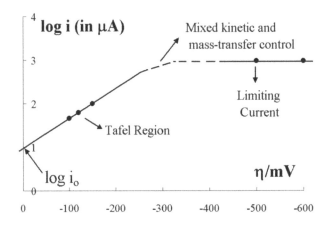

The first three points provide a Tafel line with $Slope = -6.8 = -\alpha F/2.3RT$. For T = 298 K, this yields $\alpha = 0.40$. Extrapolation to $\eta = 0$ gives $\log i_0 = 0.98$, or $i_0 = 9.5$ μA. From equation (3.4.7),

$$k^0 = \frac{i_0}{FAC} = \frac{9.5 \times 10^{-6} \text{ A}}{96485 \text{ C/mol} \times 0.1 \text{ cm}^2 \times 1 \times 10^{-5} \text{ mol/cm}^3} = 9.8 \times 10^{-5} \text{ cm/s}$$

From equation (3.4.13),

$$R_{ct} = \frac{8.31441 \text{ J mol}^{-1}\text{K}^{-1} \times 298 \text{ K}}{96485 \text{ C mol}^{-1} \times 9.5 \times 10^{-6} \text{ A}} = 2.7 \text{ k}\Omega$$

From $i_{l,c} = 965$ μA and equation (1.4.7),

$$m_0 = \frac{i_{\ell,c}}{FAC_0^*} = \frac{965 \times 10^{-6} \text{ A}}{96485 \text{ C mol}^{-1} \times 0.1 \text{ cm}^2 \times 1.0 \times 10^{-5} \text{ mol cm}^{-3}} = 0.010 \text{ cm/s}$$

According to equation (1.3.29), the mass transfer resistance for the oxidized form is

$$R_{mt} = \frac{8.31441 \text{ J mol}^{-1}\text{K}^{-1} \times 298 \text{ K}}{96485 \text{ C mol}^{-1} \times 965 \times 10^{-6} \text{ A}} = 26.6 \text{ }\Omega$$

Problem 3.10[©] From equation (3.5.27),

$$
\begin{aligned}
\lambda_o &= \frac{e^2}{8\pi\varepsilon_0}\left(\frac{1}{a_0}-\frac{1}{R}\right)\left(\frac{1}{\varepsilon_{op}}-\frac{1}{\varepsilon_s}\right) \\[2mm]
&= \frac{\left(1.60219\times10^{-19}\mathrm{C}\right)^2}{8\times\pi\times8.85419\times10^{-12}\ \mathrm{C^2N^{-1}m^{-2}}}\left[\frac{1}{7.0\times10^{-10}\ \mathrm{m}}-\frac{1}{14\times10^{-10}\ \mathrm{m}}\right]\times0.5 \\[2mm]
&= 4.12\times10^{-20}\ \mathrm{J} = 0.26\ \mathrm{eV}
\end{aligned}
$$
(3.5.27)

where λ_o is the reorganization energy of the solvent and R is $2\times7\times10^{-10}$ m. The reorganization of the electroactive species is ignored in this problem ($\lambda_i = 0$), so that from $\lambda = \lambda_i + \lambda_o$ (equation (3.5.25)).

$$\lambda = \lambda_o$$

From equation (3.5.21),

$$\Delta G^{\ddagger} = \frac{\lambda}{4}\left(1+\frac{F\left(E-E^{0\prime}\right)}{\lambda}\right)^2$$
(3.5.21)

For $E = E^0$,

$$\Delta G^{\ddagger} = \frac{\lambda}{4} = \frac{0.257\ \mathrm{eV}}{4} = 0.0643\ \mathrm{eV}$$

=====

Problem 3.12[©] Starting with equation (3.5.57) for $D_O(\lambda, \mathbf{E})$ and equation (3.5.55) for $D_R(\lambda, \mathbf{E})$, and substituting for $W_O(\lambda, \mathbf{E})$ from equation (3.5.67 and $W_R(\lambda, \mathbf{E})$ from equation (3.5.68) leads to the following expressions.

$$D_O(\lambda, \mathbf{E}) = \frac{N_A C_O(0,t)}{(4\pi\lambda kT)^{1/2}}\exp\left[-(\mathbf{E}-\mathbf{E}^0-\lambda)^2/4\lambda kT\right]$$
((3.5.57))

$$D_R(\lambda, \mathbf{E}) = \frac{N_A C_R(0,t)}{(4\pi\lambda kT)^{1/2}}\exp\left[-(\mathbf{E}-\mathbf{E}^0+\lambda)^2/4\lambda kT\right]$$
(3.5.55)

At $\mathbf{E} = \mathbf{E}_{eq}$, $C_O(0,t) = C_O^*$ and $C_R(0,t) = C_R^*$ so that equations (3.5.57) and 3.5.55 can be rewritten as

$$D_O(\lambda, \mathbf{E}_{eq}) = \frac{N_A C_O^*}{(4\pi\lambda kT)^{1/2}}\exp\left[-(\mathbf{E}_{eq}-\mathbf{E}^0-\lambda)^2/4\lambda kT\right]$$

$$D_R(\lambda, \mathbf{E}_{eq}) = \frac{N_A C_R^*}{(4\pi\lambda kT)^{1/2}}\exp\left[-(\mathbf{E}_{eq}-\mathbf{E}^0+\lambda)^2/4\lambda kT\right]$$

Chapter 3   BASIC KINETICS OF ELECTRODE REACTIONS

At equilibrium, the Fermi level must be uniform everywhere in the system. Thus, the occupancy at $\mathbf{E}_{eq}$ must be uniformly at 50%. On the solution side, this condition implies that the concentration density functions $D_O(\lambda, \mathbf{E}_{eq})$ and $D_R(\lambda, \mathbf{E}_{eq})$ are equal.

$$\frac{N_A C_O^*}{(4\pi\lambda kT)^{1/2}} \exp\left[\frac{-(\mathbf{E}_{eq} - \mathbf{E}^0 - \lambda)^2}{4\lambda kT}\right] = \frac{N_A C_R^*}{(4\pi\lambda kT)^{1/2}} \exp\left[\frac{-(\mathbf{E}_{eq} - \mathbf{E}^0 + \lambda)^2}{4\lambda kT}\right]$$

Solving for C_O^*/C_R^*, one finds

$$\frac{C_O^*}{C_R^*} = \exp\left[\frac{-(\mathbf{E}_{eq} - \mathbf{E}^0 + \lambda)^2 + (\mathbf{E}_{eq} - \mathbf{E}^0 - \lambda)^2}{4\lambda kT}\right]$$

which, after some algebra, reduces to

$$\frac{C_O^*}{C_R^*} = \exp\left[\frac{(\mathbf{E}^0 - \mathbf{E}_{eq})}{kT}\right]$$

and further to

$$\mathbf{E}_{eq} = \mathbf{E}^0 - kT \ln \frac{C_O^*}{C_R^*} \tag{1}$$

Note that equation (1) has the general form of equation (3.2.2), the Nernst equation.

$$E_{eq} = E^{0\prime} + \frac{RT}{F} \ln \frac{C_O^*}{C_R^*} \tag{2}$$

However, equation (1) is expressed in terms of electron energy.

Electrical potential E is the potential energy per unit charge. Electrical potential is measured in volts where V = J/C. Energy $\mathbf{E}$, is measured in joules. Thus, potential is energy per charge. The Nernst equation (equation (2)) is readily derived from equation (1) by dividing the energies of equation (1) by electrical charge $-e$, such that $E_{eq} = -\mathbf{E}_{eq}/e$ and $E^{0\prime} = -\mathbf{E}^0/e$.

$$E_{eq} = E^0 + \frac{kT}{e} \ln \frac{C_O^*}{C_R^*}$$

This is converted to the familiar form by multiplying both the numerator and the denominator of the second term by the Avogadro number, N_A. Recognize the per molecule fundamental constants k and e are converted to per mole constants R and F because $N_A k = R$ and $N_A e = F$. Thus, the Nernst equation (2) is derived from equation (1).

4 MASS TRANSFER BY MIGRATION AND DIFFUSION

Problem 4.2[©] The only ionic species in solution are Na^+ and OH^-, both present at 0.10 M. Using equation (4.2.9), the transference number for Na^+ is

$$t_{Na^+} = \frac{|z_{Na^+}|C_{Na^+}\lambda_{Na^+}}{|z_{Na^+}|C_{Na^+}\lambda_{Na^+} + |z_{OH^-}|C_{OH^-}\lambda_{OH^-}}$$

Because $|z_{Na^+}| = |z_{OH^-}|$ and $C_{Na^+} = C_{OH^-}$, this expression reduces to

$$t_{Na^+} = \frac{\lambda_{o,Na^+}}{\lambda_{o,Na^+} + \lambda_{o,OH^-}} = \frac{50.11}{50.11 + 198} = 0.20$$

where λ_0 has been substituted for λ. From equation (2.3.6),

$$t_{OH^-} = 1 - t_{Na^+} = 0.80$$

For 20 e passed externally, 20 e are injected at the cathode and 20 e are withdrawn at the anode. Thus, 20 OH^- are created at the cathode and 20 OH^- are removed at the anode. These changes are shown in the balance sheet below.

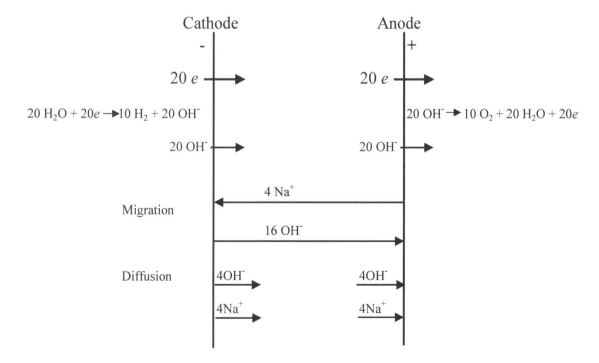

Electrochemical Methods: Fundamentals and Applications, Third Edition, Student Solutions Manual. Cynthia G. Zoski and Johna Leddy.
© 2025 John Wiley & Sons Ltd. Published 2025 by John Wiley & Sons Ltd.

Chapter 4   MASS TRANSFER BY MIGRATION AND DIFFUSION

In the bulk solution, charge is transported only by migration, a fraction 0.80 being carried by OH^- moving to the anode and a fraction 0.20 carried by Na^+ moving to the cathode. Thus, 16 OH^- and 4 Na^+ migrate through the bulk and through diffusion layers, as shown in the balance sheet. This result could also be obtained via equation (4.3.3) by considering current flow at either electrode.

At the anode 20 OH^- are consumed, 16 of which are supplied by migration. The remaining 4 must diffuse to the electrode. No Na^+ is consumed or generated, yet 4 Na^+ exit by migration. At steady state, they must be replaced by diffusion to maintain a constant concentration distribution. Likewise, the cathode generates 20 OH^- and no Na^+, while 16 OH^- leave and 4 Na^+ arrive by migration. Thus, 4 OH^- and 4 Na^+ must diffuse outward per 20 e at steady state. The fluxes from diffusion complete the balance sheet.

=====

Problem 4.4[©]   For linear diffusion, the diffusion length can be estimated from the root-mean-square displacement given as equation (4.4.3).

$$\overline{\Delta} = \sqrt{2Dt} \tag{4.4.3}$$

Thus, the minimum distance d between the working electrode surface and the cell wall is

$$d = 5\overline{\Delta} = 5\sqrt{2Dt} = 5\sqrt{2 \times 10^{-5}\ \text{cm}^2/\text{s} \times 100\ \text{s}} = 0.2\ \text{cm}$$

=====

Problem 4.5[©]   The Nernst-Einstein equation (equation (4.1.11)) expresses the diffusion coefficient D_j in terms of mobility u_j measured at infinite dilution. At 298 K,

$$
\begin{aligned}
D_j &= \frac{RT}{|z_j|F}u_j \tag{4.1.11} \\
&= 0.02569 V u_j |z_j|^{-1}
\end{aligned}
$$

(a). From the mobility data in Table 2.3.2, the diffusion coefficients are estimated as follows.

u_j and D_j for H^+, I^-, and Li^+			
Ion	z_j	u_j (cm^2/Vs)	D_j (cm^2/s)
H^+	1	3.625×10^{-3}	9.314×10^{-5}
I^-	-1	7.96×10^{-4}	2.05×10^{-5}
Li^+	1	4.010×10^{-4}	1.030×10^{-5}

(b). From equation (2.3.15),

$$\lambda_j = F u_j \tag{2.3.15}$$

Substituting in the Nernst-Einstein equation (4.1.11),

$$D_j = \frac{RT\lambda_j}{|z_j|F^2}$$

Problem 4.6© The geometry of the problem is as shown below.

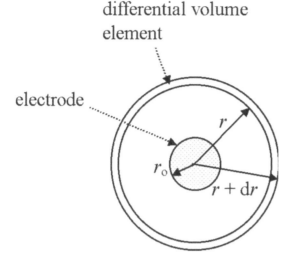

differential volume
element

electrode

Spherical symmetry implies that concentrations change along a radial line extending from the center. However, for all points at a given radius, the concentration is the same.

Net diffusive transport occurs only radially because there a gradient in concentration only in the r direction. Fick's first law, therefore, is

$$J_o(r,t) = -D_O \frac{\partial C_O(r,t)}{\partial r}$$

Now consider the volume element of thickness dr contained between the radii r and $r + dr$. The change in the number of moles of species O, N_O, within the volume element over time interval dt is the difference between the number of moles diffusing in (across boundary r) and the number diffusing out (across boundary $r + dr$). The inbound quantity is $4\pi r^2 J_O(r,t)dt$ moles, where $4\pi r^2$ is the area of the boundary at r. Likewise, the outbound quantity is $4\pi(r + dr)^2 J_O(r + dr, t)dt$. Thus, the differential change in the number of moles of O within the element is

$$dN_O = 4\pi \times dt \times \left[r^2 J_O(r,t) - (r + dr)^2 J_O(r + dr, t) \right]$$

Recognizing that

$$J_O(r + dr, t) = J_O(r,t) + \frac{\partial J_O(r,t)}{\partial r} \partial r$$

leads to

$$dN_O = 4\pi \times dt \times \left\{ \left[r^2 - (r + dr)^2 \right] J_O(r,t) - (r + dr)^2 \frac{\partial J_O(r,t)}{\partial r} \partial r \right\}$$

To obtain the change in concentration, $C_O(r,t)$, it is necessary to divide by the volume of the element, which is $4\pi r^2 dr$.

$$dC_O(r,t) = dt \times \left\{ \frac{[r^2 - (r+dr)^2]}{r^2 dr} J_O(r,t) - \frac{(r+dr)^2}{r^2} \frac{\partial J_O(r,t)}{\partial r} \right\}$$

The next step requires that $dC_O(r,t)$ be converted to the rate of change $\partial C_O(r,t)/\partial t$ by rearranging this equation, and expanding the expressions in r algebraically.

$$\frac{\partial C_O(r,t)}{\partial t} = -\left[\frac{2}{r} + \frac{dr}{r^2} \right] J_O(r,t) - \left[1 + \frac{2dr}{r} + \frac{dr^2}{r^2} \right] \frac{\partial J_O(r,t)}{\partial r}$$

Because dr is infinitesimally small, the terms containing it within the parentheses are negligible. Thus,

$$\frac{\partial C_O(r,t)}{\partial t} = -\frac{J_O(r,t)}{\partial r} - \frac{2}{r} J_O(r,t)$$

Substitution from Fick's first law then gives

$$\frac{\partial C_O(r,t)}{\partial t} = D_O \left[\frac{\partial^2 C_O(r,t)}{\partial r^2} + \frac{2}{r} \frac{\partial C_O(r,t)}{\partial r} \right] \tag{4.4.18}$$

which is equation (4.4.18).

5 STEADY-STATE VOLTAMMETRY AT ULTRAMICROELECTRODES

Problem 5.1$^{\copyright}$ The steady state current at an UME is given by equation (5.2.18b), solved for the electrode radius r_0.

$$
\begin{aligned}
r_0 &= \frac{i_{d,c}}{4nFD_OC_O^*} \qquad\qquad\qquad\qquad\qquad\qquad\qquad (5.2.18b)\\[2mm]
&= \frac{2.32 \times 10^{-9}\ \text{A}}{4 \times 1 \times 96485\ \text{C/mole} \times 1.2 \times 10^{-5}\ cm^2/s \times 1 \times 10^{-6}\ \text{mole/cm}^3}\\[2mm]
&= 5.0 \times 10^{-4}\ \text{cm} = 5.0\ \mu\text{m}
\end{aligned}
$$

=====

Problem 5.3$^{\copyright}$ In answering this question, one can follow the procedure of Section 1.3.2(c). For

$$M^{+n} + ne \rightleftharpoons M(solid)$$

the Nernst equation is specified by equation (1.3.23). Note, for a pure solid is at unit activity, $a_M = 1$.

$$
E = E^{o\prime}_{M^{n+},M(solid)} + \frac{RT}{nF} \ln C_{M^{n+}}(surface) \qquad\qquad (1)
$$

Equation (1.3.11) specifies the relationship between surface concentration, current, and limiting current.

$$
\frac{C_{M^{n+}}(surface)}{C^*_{M^{n+}}} = 1 - \frac{i}{i_d} \qquad\qquad (2)
$$

The limiting current i_d is specified at the mass transport limit in equation (1.3.10).

$$
i_d = nFAm_{M^{n+}}C^*_{M^{n+}} \qquad\qquad (3)
$$

At steady state, i_d is the limiting current. Substitution of equation (2) into equation (1) yields

$$
\begin{aligned}
E &= E^{o\prime}_{M^{n+},M(solid)} + \frac{RT}{nF} \ln C^*_{M^{n+}} \left[\frac{i_d - i}{i_d}\right]\\[2mm]
&= E^{o\prime}_{M^{n+},M(solid)} + \frac{RT}{nF} \ln \left[\frac{C^*_{M^{n+}}}{i_d}\right] + \frac{RT}{nF} \ln [i_d - i]
\end{aligned}
$$

Electrochemical Methods: Fundamentals and Applications, Third Edition, Student Solutions Manual. Cynthia G. Zoski and Johna Leddy.
© 2025 John Wiley & Sons Ltd. Published 2025 by John Wiley & Sons Ltd.

For $E = E_{1/2}$, $i = i_d/2$. From equation (3), $C^*_{M^{n+}}/i_d = [nFAm_{M^{n+}}]^{-1}$.

$$E_{1/2} = E^{o\prime}_{M^{n+},M(solid)} - \frac{RT}{nF} \ln [nFAm_{M^{n+}}] + \frac{RT}{nF} \ln \left[\frac{i_d}{2}\right] \tag{4}$$

This is the relationship for the half-wave potential $E_{1/2}$ versus the limiting current i_d. $E_{1/2}$ varies linearly with $\ln i_d$. A plot of $E_{1/2}$ versus $\ln [i_d]$ yields slope of RT/nF and intercept $E^{o\prime}_{M^{n+},M(solid)} - \frac{RT}{nF} \ln [2nFAm_{M^{n+}}]$.

Substitution of equation (3) into equation (4) yields the relationship between $E_{1/2}$ and $C^*_{M^{n+}}$.

$$
\begin{aligned}
E_{1/2} &= E^{o\prime}_{M^{n+},M(solid)} - \frac{RT}{nF} \ln [nFAm_{M^{n+}}] + \frac{RT}{nF} \ln \left[\frac{nFAm_{M^{n+}} C^*_{M^{n+}}}{2}\right] \\
&= E^{o\prime}_{M^{n+},M(solid)} + \frac{RT}{nF} \ln \left[\frac{C^*_{M^{n+}}}{2}\right] \\
&= E^{o\prime}_{M^{n+},M(solid)} - \frac{RT}{nF} \ln [2] + \frac{RT}{nF} \ln [C^*_{M^{n+}}]
\end{aligned}
$$

$E_{1/2}$ varies linearly with $\ln C^*_{M^{n+}}$. A plot of $E_{1/2}$ versus $\ln C^*_{M^{n+}}$ yields slope RT/nF and intercept $E^{o\prime}_{M^{n+},M(solid)} - 0.693\frac{RT}{nF}$.

=====

Problem 5.7[©] **(a).** Equation (5.1.20) applies for a spherical electrode, but for a hemisphere embedded in a semi-infinite insulating plane, half the current is generated. The steady-state current is used to find the diffusion coefficient given a radius of 5.0×10^{-4} cm, concentration of 1.0×10^{-5} mol/cm^3, and $n = 1$.

$$
\begin{aligned}
D_O &= \frac{i_{d,c}}{2\pi nFC^*_O r_0} \\
&= \frac{1.5 \times 10^{-8}\ A}{2 \times \pi \times 1 \,(96485\ \text{C/mol})\,(1.0 \times 10^{-5}\ \text{mol/cm}^3)\,(5.0 \times 10^{-4}\ \text{cm})} \\
&= 5.0 \times 10^{-6}\ \text{cm/s}
\end{aligned}
$$

(b). In *Anal. Chem.* **64** 2293 (1992), tables are provided for determining the standard heterogeneous rate constant and transfer coefficient from $\Delta E_{3/4}$ and $\Delta E_{1/4}$. For the values of $\Delta E_{3/4} = 35.0$ mV and $\Delta E_{1/4} = 31.5$ mV, values of $\alpha = 0.38$ and $\lambda = k^0/m_O = 3.95$ are found, where $m_O = D_O/r_0 = \left(5.0 \times 10^{-6}\ \text{cm}^2/\text{s}\right)/5.0 \times 10^{-4}$ cm $= 0.01$ cm/s. Thus, $k^0 = 0.0395$ cm/s.

Problem 5.9 ©   **(a)** Begin with equation (5.4.7).

$$i = \frac{\Lambda^0 \theta^{-\alpha} (i_{d,c} + \xi\theta i_{d.a})}{1 + \Lambda^0 \theta^{-\alpha} (1 + \xi\theta)} \tag{5.4.7}$$

Equation (5.4.7) is limited to reversible kinetics where Λ^0 is very large.

$$i = \lim_{\Lambda^0 \to \infty} \frac{\Lambda^0 \theta^{-\alpha} (i_{d,c} + \xi\theta i_{d.a})}{1 + \Lambda^0 \theta^{-\alpha} (1 + \xi\theta)} = \frac{i_{d,c} + \xi\theta i_{d.a}}{1 + \xi\theta} \tag{1}$$

Solve for $\xi\theta$.

$$\xi\theta = \frac{i_{d,c} - i}{i - i_{d,a}}$$

Equation (5.4.7) is derived for a one step, one electron reaction; $n = 1$ and $\theta = e^{f(E - E^{0\prime})}$.

$$\xi e^{f(E - E^{0\prime})} = \frac{i_{d,c} - i}{i - i_{d,a}}$$

The logarithm of both sides yields

$$\ln \xi + f\left(E - E^{0\prime}\right) = \ln \left[\frac{i_{d,c} - i}{i - i_{d,a}} \right]$$

On rearranging,

$$E = E^{0\prime} - \frac{1}{f} \ln \xi + \frac{1}{f} \ln \left[\frac{i_{d,c} - i}{i - i_{d,a}} \right]$$

Substitute $f = F/RT$ and $\xi = m_O/m_R$.

$$E = E^{0\prime} + \frac{RT}{F} \ln \frac{m_R}{m_O} + \frac{RT}{F} \ln \left[\frac{i_{d,c} - i}{i - i_{d,a}} \right]$$

From equation (5.3.5), $E_{1/2} = E^{0\prime} + \frac{RT}{F} \ln \frac{m_R}{m_O}$. This yields equation (5.3.7) for $n = 1$.

$$E = E_{1/2} + \frac{RT}{F} \ln \left[\frac{i_{d,c} - i}{i - i_{d,a}} \right] \tag{5.3.7}$$

(b) Because equation (5.3.7) is shown to be a form of equation (5.4.7) for the special case of reversible kinetics for $n = 1$, it also applies to the present problem. However, it must be restricted further to the case of $C_R^* = 0$. When this is true, $i_{d,a} = 0$. Therefore, equation (5.3.7) becomes equation (5.3.4) for $n = 1$.

$$E = E_{1/2} + \frac{RT}{F} \ln \left[\frac{i_{d,c} - i}{i} \right] \tag{5.3.4}$$

(c) The system is reversible and $n = 1$. Hence, equation (1) applies. For the case of $C_O^* = 0$ and $i_{d,c} = 0$, equation (1) becomes

$$i = \frac{\xi \theta i_{d,a}}{1 + \xi \theta} \tag{5.11.1}$$

The result is equation (5.11.1).

=====

Problem 5.11[©]   The electrode process is the reduction of A^z by one electron to produce A^{z-1}. A Pt disk UME with $r_0 = 10$ μm was employed, and the electroactive species was present at 1 mM. There are four different cases, all involving measurements of the limiting current at $\gamma = 500$ (0.5 M supporting electrolyte) and at $\gamma = 0$ (no added supporting electrolyte). From equation (5.7.3), $\gamma = C_{electrolyte}^* / C_{redox}^*$.

The two relevant equations are (5.7.4) and (5.7.5), applying when $n = z$ and $n \neq z$, respectively.

$$\frac{i_l (\gamma = 0)}{i_l (\gamma \to \infty)} = 1 + |n| \qquad \text{for } n = z \tag{5.7.4}$$

$$\frac{i_l (\gamma = 0)}{i_l (\gamma \to \infty)} = 1 \pm z \left\{ 1 + (1 + |z|) \left(1 - \frac{z}{n} \right) \ln \left[1 - \frac{1}{(1 + |z|) \left(1 - \frac{z}{n} \right)} \right] \right\} \qquad \text{for } n \neq z \tag{5.7.5}$$

In these equations, n is positive for a reduction and negative for an oxidation. Here, A^z is reduced, so n is positive in all present cases. In equation (5.7.5), the negative sign is used for $n > z$ and the positive sign for $n < z$.

(a) From the data given, $i_l (\gamma = 0) / i_l (\gamma = 500) = 1.27$. Because $i_l (\gamma = 0) / i_l (\gamma = 500) > 1$, the current $i_l (\gamma = 0)$ is enhanced by migration for reduction; the electroreactant is positively charged and attracted to the negatively polarized electrode. The ratio does not approach 2, so $z \neq +1$. More likely, $z = +2$. On substitution of $n = +1$ and $z = +2$ into equation (5.7.5), the result is

$$\frac{i_l (\gamma = 0)}{i_l (\gamma \to \infty)} = 1 + 2 \left\{ 1 + 3 (-1) \ln \left[1 - \frac{1}{3 (-1)} \right] \right\} = 3 - 6 \ln (1.33) = 1.29$$

This is in good agreement with the observed ratio of 1.27. In Figure 5.7.4, the low-γ asymptote for $z = +2$ is at about 1.3, also in good agreement. The result confirms that $i_l (\gamma = 500)$ is an appropriate approximation of $i_l (\gamma \to \infty)$. The charge on the electroreactant is then +2.

6 TRANSIENT METHODS BASED ON POTENTIAL STEPS

Problem 6.1[©] Fick's second law for diffusion to a spherical electrode of radius r_0 for species O is shown in equation (6.1.15). The equation applies to any diffusing species. The form of the equation is the same for O and for R. For either O or R,

$$\frac{\partial C(r,t)}{\partial t} = D\left[\frac{\partial^2 C(r,t)}{\partial r^2} + \frac{2}{r}\frac{\partial C(r,t)}{\partial r}\right] \tag{6.1.15}$$

This equation can be solved for $C(r,t)$ according to the boundary conditions

$$C(r,0) = C^* \qquad r > r_0 \tag{6.1.16}$$

$$\lim_{r\to\infty} C(r,t) = C^* \tag{6.1.17}$$

$$C(r_0,t) = 0 \qquad t > 0 \tag{6.1.18}$$

and making the following substitution

$$v(r,t) = rC(r,t) \tag{1}$$

Thus,

$$C(r,t) = \frac{v(r,t)}{r} \tag{2}$$

Differentiating equation (2) with respect to t and r yields the following:

$$\frac{\partial C(r,t)}{\partial t} = \frac{1}{r}\frac{\partial v(r,t)}{\partial t} \tag{3}$$

$$\frac{\partial C(r,t)}{\partial r} = \frac{1}{r}\frac{\partial v(r,t)}{\partial r} - \frac{v(r,t)}{r^2} \tag{4}$$

$$\frac{\partial^2 C(r,t)}{\partial r^2} = \frac{1}{r}\frac{\partial^2 v(r,t)}{\partial r^2} - \frac{2}{r^2}\frac{\partial v(r,t)}{\partial r} + \frac{2}{r^3}v(r,t) \tag{5}$$

Substitution of equations (3) to (5) into equation 6.1.15 leads to

$$\frac{\partial v(r,t)}{\partial t} = D\frac{\partial^2 v(r,t)}{\partial r^2} \tag{6}$$

which now has the form of the linear diffusion equation so that one can proceed as in Section 6.1.1(a) for a planar electrode. Writing the boundary conditions given by equations (6.1.16) - (6.1.18) in terms of $v(r,t)$ one obtains

$$v(r,0) = rC^* \tag{7}$$

$$v(r_o,t) = 0 \qquad\qquad t > 0 \tag{8}$$

$$\lim_{r\to\infty} v(r,t) = rC^* \tag{9}$$

Electrochemical Methods: Fundamentals and Applications, Third Edition, Student Solutions Manual. Cynthia G. Zoski and Johna Leddy.
© 2025 John Wiley & Sons Ltd. Published 2025 by John Wiley & Sons Ltd.

Chapter 6   TRANSIENT METHODS BASED ON POTENTIAL STEPS

The Laplace transform of equation (6), from A.1.6, and with the boundary condition given by equation (7) leads to

$$s\overline{v}(r,s) - v(r,0) = D\frac{\partial^2 v(r,s)}{\partial r^2} \tag{10}$$

Rearrangement leads to

$$\frac{\partial^2 \overline{v}(r,s)}{\partial r^2} - \frac{s}{D}\overline{v}(r,s) = -\frac{v(r,0)}{D} \tag{11}$$

which has the solution

$$\overline{v}(r,s) = \frac{v(r,0)}{s} + A'(s)\exp\left[-\sqrt{\frac{s}{D}}r\right] + B'(s)\exp\left[\sqrt{\frac{s}{D}}r\right] \tag{12}$$

Considering the boundary condition given by equation (9), allows the determination of the constant $B'(s)$ as follows:

$$\lim_{r\to\infty} v(r,t) = rC^* = v(r,0) \tag{13}$$

$$\overline{v}(r,s) = \frac{v(r,0)}{s} \tag{14}$$

Thus, as $r \to \infty$, $B'(s)$ in equation (12) must go to zero for the solution to be true (since otherwise the exponential term goes to infinity). The $A'(s)$ constant can be determined by incorporation of the boundary condition given by equation (8), which, when Laplace transformed becomes

$$\overline{v}(r_0,s) = 0 \tag{15}$$

so that equation (12) can be written

$$\overline{v}(r_0,s) = 0 = \frac{v(r_0,0)}{s} + A'(s)\exp\left[-\sqrt{\frac{s}{D}}r_0\right] \tag{16}$$

Solving for $A'(s)$ leads to

$$A'(s) = -\frac{v(r_0,0)}{s}\exp\left[\sqrt{\frac{s}{D}}r_0\right] \tag{17}$$

Substitution for $A'(s)$ and $B'(s)$ into equation (12) leads to

$$\overline{v}(r,s) = \frac{v(r,0)}{s} - \frac{v(r_0,0)}{s}\exp\left[-\sqrt{\frac{s}{D}}(r - r_0)\right] \tag{18}$$

One now resubstitutes for the $\overline{v}(r,s)$ terms by making use of equation (7) and the Laplace transform of equation (1)

$$\overline{v}(r,s) = r\overline{C}(r,s) \tag{19}$$

to yield

$$\overline{C}(r, s) = \frac{C^*}{s} - \frac{r_0}{rs} C^* \exp\left[-\sqrt{\frac{s}{D}}(r - r_0)\right] \tag{20}$$

The flux at a spherical electrode surface follows from equation (4.5.7).

$$-J(0, t) = \frac{i(t)}{nFA} = D\left[\frac{\partial C_O(r, t)}{\partial r}\right]_{r=r_0} \tag{4.5.7}$$

which Laplace transforms to

$$\frac{\bar{i}(s)}{nFA} = D\left[\frac{\partial \overline{C}(r, s)}{\partial r}\right]_{r=r_0} \tag{21}$$

Taking the partial derivative of $\bar{C}(r, s)$ with respect to r leads to

$$\frac{\partial \overline{C}(r, s)}{\partial r} = \frac{r_0}{sr^2} C^* \exp\left[-\sqrt{\frac{s}{D}}(r - r_0)\right] + \frac{r_0}{r}\frac{C^*}{\sqrt{sD}} \exp\left[-\sqrt{\frac{s}{D}}(r - r_0)\right]$$

and specifying the condition $r = r_0$ results in

$$\left[\frac{\partial \overline{C}(r, s)}{\partial r}\right]_{r=r_0} = \frac{C^*}{sr_0} + \frac{C^*}{\sqrt{sD}} \tag{22}$$

Substitution of equation (22) into equation (21) leads to

$$\bar{i}(s) = nFADC^*\left[\frac{1}{sr_0} + \frac{1}{\sqrt{sD}}\right] \tag{23}$$

From Table A.1.1, $f(s) = s^{-1/2}$ inverse Laplace transforms to $F(t) = (\pi t)^{-1/2}$ and A/s to A(constant). Thus, the inverse Laplace transform of equation (23) is

$$i(t) = nFADC^*\left[\frac{1}{r_0} + \frac{1}{\sqrt{\pi Dt}}\right] \tag{24}$$

This result is the same as equation (6.1.19) where reactant O is specified.

$$i(t) = nFAD_O C_O^*\left[\frac{1}{\sqrt{\pi D_O t}} + \frac{1}{r_0}\right] \tag{6.1.19}$$

Chapter 6   TRANSIENT METHODS BASED ON POTENTIAL STEPS

Problem 6.2©   The equations for planar and spherical diffusion are similar. For an electroreactant in a quiescent semi-infinite system with a uniform concentration C^* at $t = 0$,

$$i(t) = nFADC^* \left[\frac{1}{\sqrt{\pi D t}} + \frac{b}{r_0} \right]$$

where $b = 0$ for a planar electrode (as in equation 6.1.12) and $b = 1$ for a spherical electrode (as in equation 6.1.19). The currents for each are shown in the table below. Note that for $A = 0.02$ cm^2 $= 4\pi r_0^2$, the radius of the spherical electrode is $r_0 = 0.040$ cm.

n	1	
C*	1.00E-06	mol/cm^3
A	0.02	cm^2
D	1.00E-05	cm^2/s
F	96485	C/mol
nFAC*D	1.93E-08	Acm
r$_0$	0.04	cm

t(s)	$i_{planar}(\mu A)$	$i_{spherical}(\mu A)$
0.1	10.89	11.37
0.5	4.87	5.35
1	3.44	3.93
2	2.43	2.92
3	1.99	2.47
5	1.54	2.02
10	1.09	1.57
20	0.77	1.25
30	0.63	1.11
inf	0.00	0.48

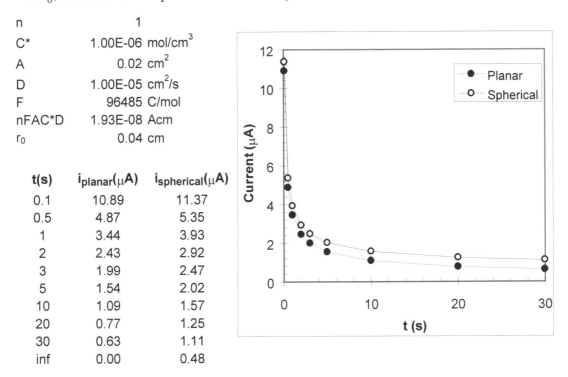

The electrolysis at the spherical electrode exceeds that at the planar electrode by 10% when

$$\frac{i_{spherical}}{i_{planar}} = \frac{\frac{1}{\sqrt{\pi D t}} + \frac{1}{r_0}}{\frac{1}{\sqrt{\pi D t}}} > 1.1$$

$$\frac{1}{r_0} > 0.1 \frac{1}{\sqrt{\pi D t}}$$

$$10\sqrt{\pi D t} > r_0$$

$$t > \frac{r_0^2}{10^2 \pi D}$$

For this system, this corresponds to

$$t > \frac{(0.04)^2}{100\pi \times 10^{-5}} = 0.51 \text{ s}$$

This is consistent with the values shown in the spreadsheet for $t = 0.5$ s.

Cottrell's equation (equation (6.1.12)) is

$$i(t) = \frac{nFAC^*\sqrt{D}}{\sqrt{\pi t}} \tag{6.1.12}$$

This integrates with respect to t to yield the charge.

$$Q_d(t) = \frac{2nFAC^*\sqrt{Dt}}{\sqrt{\pi}} \tag{6.6.1}$$

For $t = 10$ s and the values listed above, $Q_d(t) = 2.2 \times 10^{-5}$ C. From Faraday's Law, $Q/nF =$ moles electrolyzed yields 2.3×10^{-10} moles. In 10 mL of 1 mM solution, there are 10^{-5} moles of material. In 10 s, the fraction electrolyzed is $2.3 \times 10^{-10}/10^{-5}$ or 0.0023% . Thus, under conditions for normal voltammetric measurements, the bulk concentration of the redox species is not perturbed significantly.

$$===== $$

Problem 6.4$^{©}$ To derive the Nernst expression for

$$MX_p^{2-p} + 2e + Hg \rightleftharpoons M(Hg) + pX^- \qquad E_C^{0\prime}$$

begin with the reaction to form the metal amalgam.

$$M^{2+} + 2e + Hg \rightleftharpoons M(Hg) \qquad E_M^{0\prime}$$

The corresponding Nernst equation is

$$E = E_M^{0\prime} - \frac{RT}{2F} \ln \frac{C_{M(Hg)}(0,t)}{C_{M^{2+}}(0,t)} \tag{1}$$

Complexation $M^{2+} + pX^- \rightleftharpoons MX_p^{2-p}$ is expressed in the complexation or formation constant.

$$K_C = \frac{C_{MX_p^{2-p}}}{C_{M^{2+}}C_{X^-}^p} \tag{2}$$

Solve for $C_{M^{2+}}$ and substitute into equation (1).

$$E = E_M^{0\prime} - \frac{RT}{2F} \ln \frac{C_{M(Hg)}(0,t)K_C C_{X^-}^p(0,t)}{C_{MX_p^{2-p}}(0,t)}$$

Chapter 6   TRANSIENT METHODS BASED ON POTENTIAL STEPS

This rearranges to the Nernst expression for the 2-electron reduction of MX_p^{2-p}.

$$E = E_M^{0\prime} - \frac{RT}{2F} \ln K_C - \frac{pRT}{2F} \ln C_{X^-}(0,t) - \frac{RT}{2F} \ln \frac{C_{M(Hg)}(0,t)}{C_{MX_p^{2-p}}(0,t)}$$

Consider conditions where the initial concentration of the amalgam is zero; the bulk concentrations of MX_p^{2-p} and X^- are constant at $C_{MX_p^{2-p}}^*$ and $C_{X^-}^*$ and because $C_{X^-}^* >> C_{MX_p^{2-p}}^*$, $C_{X^-}(0,t) \longrightarrow C_{X^-}^*$.

$$E = E_M^{0\prime} - \frac{RT}{2F} \ln K_C - \frac{pRT}{2F} \ln C_{X^-}^* - \frac{RT}{2F} \ln \frac{C_{M(Hg)}(0,t)}{C_{MX_p^{2-p}}(0,t)} \tag{3}$$

For a reversible sampled transient voltammogram, the simplified current-concentration relationships in Section 6.2.4 apply. For m_C and m_A the mass transport coefficients (cm/s) for the complex and the amalgam, the current $i(t)$ at time t for the complex, the metal amalgam are defined.

$$
\begin{aligned}
i(t) &= 2FAm_C \left[C_{MX_p^{2-p}}^* - C_{MX_p^{2-p}}(0,t) \right] \\
i(t) &= 2FAm_A C_{M(Hg)}(0.t)
\end{aligned}
$$

The diffusion limited current for the complex $i_d(t)$ is for $C_{MX_p^{2-p}}(0,t) = 0$ and E is well negative of $E_C^{0\prime}$.

$$i_d(t) = 2FAm_C C_{MX_p^{2-p}}^*$$

Expressions for $C_{MX_p^{2-p}}(0,t)$ and $C_{M(Hg)}(0.t)$ are found for a given t on rearranging.

$$
\begin{aligned}
C_{MX_p^{2-p}}(0,t) &= \frac{i_d(t) - i(t)}{2FAm_C} \\
C_{M(Hg)}(0.t) &= \frac{i(t)}{2FAm_A}
\end{aligned}
$$

Substitution into equation (3) yields the expression for the complex.

$$
\begin{aligned}
E &= E_M^{0\prime} - \frac{RT}{2F} \ln K_C - \frac{pRT}{2F} \ln C_{X^-}^* - \frac{RT}{2F} \ln \frac{\frac{i(t)}{2FAm_A}}{\frac{i_d(t)-i(t)}{2FAm_C}} \\
&= E_M^{0\prime} - \frac{RT}{2F} \ln K_C - \frac{pRT}{2F} \ln C_{X^-}^* - \frac{RT}{2F} \ln \frac{m_C}{m_A} - \frac{RT}{2F} \ln \frac{i(t)}{i_d(t) - i(t)}
\end{aligned}
$$

The half wave potential $E_{1/2}^C$ is defined where $i(t) = i_d(t)/2$.

$$E_{1/2}^C = E_M^{0\prime} - \frac{RT}{2F} \ln K_C - \frac{pRT}{2F} \ln C_{X^-}^* + \frac{RT}{2F} \ln \frac{m_A}{m_C} \tag{4}$$

(a). Given equation (4), a plot of $E_{1/2}^C$ versus $\ln C_X^*$ yields a slope of $-pRT/2F$. The intercept is equal to $E_M^{0'} - \frac{RT}{2F} \ln K_C + \frac{RT}{2F} \ln \frac{m_A}{m_C}$. Linear regression of the data yields $E_{1/2}^C = -0.0513 \ln C_X^* - 0.566$ with $r = 0.99998$. Thus, $-p = 2 \times 38.92 \text{ V}^{-1} \times -0.0513 = -3.99$ or p is 4.

(b). From equation (2), $K_C = C_{MX4}/C_M C_X^4$ is the formation or stability constant for the reaction $M^{2+} + 4X^- \rightleftharpoons MX_4^{2-}$. From equation (4) and evaluation of the half wave potential for the metal amalgam in equation (1) with no ligand present, $E_{1/2}^M$, one can solve for K_C. Under the assumption that all mass transfer coefficients are equal,

$$K_C = \exp\left[-\frac{2F}{RT}(E_{1/2}^C - E_{1/2}^M) - p \ln C_X^* \right] \tag{5}$$

n	2
p	4
F/RT(V^{-1})	38.92
E$_{1/2,M}$(V)	0.081

E$_{1/2,C}$(V)	nF(E$_{1/2,C}$-E$_{1/2,M}$)/RT	C$_X$*	ln(C$_X$*)	pln(C$_X$*)	K$_C$
-0.448	-41.17736	0.10	-2.30259	-9.21034	7.64012E+21
-0.531	-47.63808	0.50	-0.69315	-2.77259	7.81763E+21
-0.566	-50.36248	1.00	0	0	7.44984E+21

An average of the last column leads to a stability constant of 7.6×10^{21}. Alternatively, from equation (5), a plot of $-nF(E_{1/2}^C - E_{1/2}^M)/RT$ versus $\ln C_X^*$ leads to a slope of p and an intercept of $\ln K_C$ when $m_A = m_C$. A linear regression of the data given leads to $p = 3.993 = 4$ and $\ln K_C = 50.38$ (with $r = 0.99998$) or $K_C = 7.58 \times 10^{21} = 7.6 \times 10^{21}$, which agrees with the previous result.

$$=====$$

Problem 6.5[C]   Consider the generic case of linear diffusion to a planar electrode. This is specified by Fick's second law (1), the initial condition (2), and the semi-infinite boundary condition (3). Fick's second law, expressed in equation (6.1.2), is applicable to both O and R.

$$\frac{\partial C(x,t)}{\partial t} = D\frac{\partial^2 C(x,t)}{\partial x^2} \tag{1}$$

$$C(x,0) = g \tag{2}$$

$$\lim_{x \to \infty} C(x,t) = g \tag{3}$$

The Laplace transform of equation (1) with respect to t yields

$$s\overline{C}(x,s) - C(x,0) = D\frac{\partial^2 \overline{C}(x,s)}{\partial x^2} \tag{4}$$

From Appendix A equation (A.1.32), this is an equation of the form

$$\frac{\partial^2 y(x)}{\partial x^2} - a^2 y(x) + b = 0 \tag{A.1.32}$$

Chapter 6   TRANSIENT METHODS BASED ON POTENTIAL STEPS

This has a solution of the form of Equation (A.1.40).

$$y(x) = \frac{b}{a^2} + A(s) \exp\left[-ax\right] + B(s) \exp\left[ax\right] \qquad \text{(A.1.40)}$$

Thus, equation (4) becomes

$$\overline{C}\left(x, s\right) = \frac{g}{s} + A(s) \exp\left[-\sqrt{\frac{s}{D}}x\right] + B(s) \exp\left[\sqrt{\frac{s}{D}}x\right] \qquad \text{(5)}$$

This is the generic solution for linear diffusion. If the system is a semi-infinite system, as characterized by equation (3), then $\overline{C}(x, s)$ must be bounded as $x \to \infty$, and, thus, $B(s) = 0$. The generic solution for linear diffusion under semi-infinite conditions is

$$\overline{C}\left(x, s\right) = \frac{g}{s} + A(s) \exp\left[-\sqrt{\frac{s}{D}}x\right] \qquad \text{(6)}$$

When both O and R are present in solution, it is necessary to develop the diffusion equations and boundary and initial conditions for both species, unless the conditions are either mass transport limited or irreversible electrode kinetics.

$$\frac{\partial C_O(x,t)}{\partial t} = D_O \frac{\partial^2 C_O\left(x, t\right)}{\partial x^2} \qquad \text{(6.1.12)}$$

$$\frac{\partial C_R(x,t)}{\partial t} = D_R \frac{\partial^2 C_R\left(x, t\right)}{\partial x^2} \qquad \text{(7)}$$

$$C_O\left(x, 0\right) = C_O^* \qquad \text{(6.1.3)}$$

$$C_R\left(x, 0\right) = C_R^* \qquad \text{(8)}$$

$$\lim_{x \to \infty} C_O\left(x, t\right) = C_O^* \qquad \text{(6.1.4)}$$

$$\lim_{x \to \infty} C_R\left(x, t\right) = C_R^* \qquad \text{(9)}$$

$$D_O \left.\frac{\partial C_O(x,t)}{\partial x}\right|_{x=0} = -D_R \left.\frac{\partial C_R(x,t)}{\partial x}\right|_{x=0} \qquad \text{(10)}$$

$$\theta = \frac{C_O(0,t)}{C_R(0,t)} = \exp\left[\frac{nF}{RT}\left(E - E^{0'}\right)\right] \qquad \text{(6.2.4)}$$

$$\frac{i(t)}{nFA} = D_O \left.\frac{\partial C_O(x,t)}{\partial x}\right|_{x=0} \qquad \text{(6.1.9)}$$

The Laplace transform of equations 6.1.12 and 6.1.12 under conditions of equations 6.1.3 to 6.1.4 yields expressions of the form of equation (6).

$$\overline{C}_O(x, s) = \frac{C_O^*}{s} + A(s) \exp\left[-\sqrt{\frac{s}{D_O}}x\right] \tag{6.2.2}$$

$$\overline{C}_R(x, s) = \frac{C_R^*}{s} + G(s) \exp\left[-\sqrt{\frac{s}{D_R}}x\right] \tag{11}$$

The Laplace transform of equations (10), 6.2.4, and 6.1.9 yields the following.

$$D_O \frac{\partial \overline{C}_O(x, s)}{\partial x}\bigg|_{x=0} = -D_R \frac{\partial \overline{C}_R(x, s)}{\partial x}\bigg|_{x=0} \tag{12}$$

$$\overline{C}_O(0, s) = \theta \overline{C}_R(0, s) \tag{13}$$

$$\frac{\overline{i}(s)}{nFA} = D_O \frac{\partial \overline{C}_O(x, s)}{\partial x}\bigg|_{x=0} \tag{14}$$

From equation (12), and given $\xi^2 = D_O/D_R$,

$$-\sqrt{sD_O}A(s) = \sqrt{sD_R}G(s)$$

Or,

$$G(s) = -\xi A(s) \tag{15}$$

From the above equation and equation (13),

$$\frac{C_O^*}{s} + A(s) = \theta\left(\frac{C_R^*}{s} + G(s)\right) = \theta\left(\frac{C_R^*}{s} - \xi A(s)\right)$$

Upon rearranging, $A(s)$ is found.

$$A(s) = -\frac{C_O^* - \theta C_R^*}{s(1 + \xi\theta)}$$

Substituting into equations 6.2.2 and 6.2.2 yields the following:

$$\overline{C}_O(x, s) = \frac{C_O^*}{s} - \frac{[C_O^* - \theta C_R^*]}{s(1 + \xi\theta)} \exp\left[-\sqrt{\frac{s}{D_O}}x\right] \tag{16}$$

$$\overline{C}_R(x, s) = \frac{C_R^*}{s} + \frac{\xi[C_O^* - \theta C_R^*]}{s(1 + \xi\theta)} \exp\left[-\sqrt{\frac{s}{D_R}}x\right] \tag{17}$$

Chapter 6   TRANSIENT METHODS BASED ON POTENTIAL STEPS

The current is found from equation (14).

$$\frac{\bar{i}(s)}{nFA} = D_O \left.\frac{\partial \overline{C}_O(x,s)}{\partial x}\right|_{x=0} = \sqrt{\frac{D_O}{s}}\frac{[C_O^* - \theta C_R^*]}{(1+\xi\theta)} \tag{18}$$

This is inverted using $s^{-1/2} \Leftrightarrow (\pi t)^{-1/2}$.

$$\frac{i(t)}{nFA\sqrt{D_O}} = \frac{1}{\sqrt{\pi t}}\frac{[C_O^* - \theta C_R^*]}{(1+\xi\theta)} \tag{19}$$

A sampled current voltammogram is the current at a time τ from a series of potential steps made to various values of E, where $\theta = C_O(0,t)/C_R(0,t) = \exp\left[nF\left(E - E^{0'}\right)/RT\right]$. When O and R are both present in solution, there will be two diffusion limited currents, one for reduction $(i_{d,c}(\tau) = nFAC_O^*\sqrt{D_O/\pi\tau})$ and one for oxidation $(i_{d,a}(\tau) = -nFAC_R^*\sqrt{D_R/\pi\tau})$.

Rearrange equation (19),

$$i(\tau) = \frac{nFA\sqrt{D_O}C_O^*}{\sqrt{\pi t}}\frac{\left[1-\theta\frac{C_R^*}{C_O^*}\right]}{(1+\xi\theta)} = i_{d,c}(\tau)\left[\frac{1-\theta\frac{C_R^*}{C_O^*}}{1+\xi\theta}\right]$$

Divide both sides by the current magnitude $i_{d,c}-i_{d,a}$. Then, note that $i_{d,a}(\tau)/i_{d,c}(\tau) = -C_R^*/\xi C_O^*$.

$$\frac{i(\tau)}{i_{d,c}(\tau)-i_{d,a}(\tau)} = \frac{i_{d,c}(\tau)}{i_{d,c}(\tau)-i_{d,a}(\tau)}\left[\frac{1-\theta\frac{C_R^*}{C_O^*}}{1+\xi\theta}\right] = \frac{1}{\left(1-\frac{i_{d,a}(\tau)}{i_{d,c}(\tau)}\right)}\left[\frac{1-\theta\frac{C_R^*}{C_O^*}}{1+\xi\theta}\right] \tag{20}$$

$$= \frac{1}{\left(1+\frac{C_R^*}{\xi C_O^*}\right)}\left[\frac{1-\theta\frac{C_R^*}{C_O^*}}{1+\xi\theta}\right]$$

From equation (20), the dimensionless ratio $X(\tau) = i(\tau)\left[i_{d,c}(\tau)-i_{d,a}(\tau)\right]^{-1}$ for the sampled current voltammogram is plotted in the spreadsheet on page 51 for several values of $a = C_R^*/C_O^*$ and ξ. (ξ is noted in the figure legend as y.) The half wave potential arises at the potential corresponding to the midpoint between $i_{d,c}$ and $i_{d,a}$. This point is darkened for each curve in the Figure. When $\xi (= y)$ is 1, the half wave potentials fall at $E = E^{0'}$, independent of the concentration ratio. These curves are denoted by a line through each set of data. When $\xi \neq 1$, the half wave potentials are at values of E different from $E^{0'}$. For $\xi < 1$, the half wave potential is shifted positive; for $\xi > 1$, the shift is negative. The concentration ratio shifts the curve up and down on the y-axis but not along the potential axis. This is a dependence of the half wave potential on ξ but not the concentration ratio. The relative magnitudes of the limiting current scale varies with concentration and ξ.

The analytical current-potential relationship is found by expressing equation (19) in terms of currents, and solving for θ. From equation (20),

$$i(\tau) = \frac{i_{d,c}(\tau) + \theta\xi i_{d,a}(\tau)}{1+\xi\theta}$$

$$\theta = \exp\left[\frac{nF}{RT}\left(E - E^{0'}\right)\right] = \frac{i_{d,c}(\tau) - i(\tau)}{\xi\left[i(\tau) - i_{d,a}(\tau)\right]}$$

Take the natural logarithm of both sides and solve for E^0. Note that $E_{1/2} = E^{0'} - \frac{RT}{nF} \ln \xi$ (equation 6.2.13). This is equation (6.2.16).

$$
\begin{aligned}
E &= E^{0'} - \frac{RT}{nF} \ln \xi + \frac{RT}{nF} \ln \left[\frac{i_{d,c}(\tau) - i(\tau)}{i(\tau) - i_{d,a}(\tau)} \right] \\
&= E_{1/2} + \frac{RT}{nF} \ln \left[\frac{i_{d,c}(\tau) - i(\tau)}{i(\tau) - i_{d,a}(\tau)} \right] \quad (21)
\end{aligned}
$$

The rightmost term goes to zero when $i(\tau) = (i_{d,c} + i_{d,a})/2$. This leaves $E_{1/2}$, which is independent of concentration but exhibits the usual dependence on ξ in equation (6.2.13).

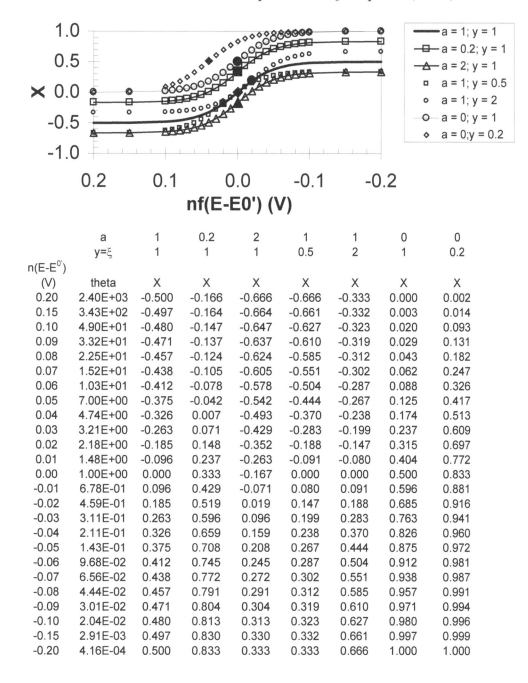

n(E-E0')		a = 1	a = 0.2	a = 2	a = 1	a = 1	a = 0	a = 0
		y=ξ = 1	1	1	0.5	2	1	0.2
(V)	theta	X	X	X	X	X	X	X
0.20	2.40E+03	-0.500	-0.166	-0.666	-0.666	-0.333	0.000	0.002
0.15	3.43E+02	-0.497	-0.164	-0.664	-0.661	-0.332	0.003	0.014
0.10	4.90E+01	-0.480	-0.147	-0.647	-0.627	-0.323	0.020	0.093
0.09	3.32E+01	-0.471	-0.137	-0.637	-0.610	-0.319	0.029	0.131
0.08	2.25E+01	-0.457	-0.124	-0.624	-0.585	-0.312	0.043	0.182
0.07	1.52E+01	-0.438	-0.105	-0.605	-0.551	-0.302	0.062	0.247
0.06	1.03E+01	-0.412	-0.078	-0.578	-0.504	-0.287	0.088	0.326
0.05	7.00E+00	-0.375	-0.042	-0.542	-0.444	-0.267	0.125	0.417
0.04	4.74E+00	-0.326	0.007	-0.493	-0.370	-0.238	0.174	0.513
0.03	3.21E+00	-0.263	0.071	-0.429	-0.283	-0.199	0.237	0.609
0.02	2.18E+00	-0.185	0.148	-0.352	-0.188	-0.147	0.315	0.697
0.01	1.48E+00	-0.096	0.237	-0.263	-0.091	-0.080	0.404	0.772
0.00	1.00E+00	0.000	0.333	-0.167	0.000	0.000	0.500	0.833
-0.01	6.78E-01	0.096	0.429	-0.071	0.080	0.091	0.596	0.881
-0.02	4.59E-01	0.185	0.519	0.019	0.147	0.188	0.685	0.916
-0.03	3.11E-01	0.263	0.596	0.096	0.199	0.283	0.763	0.941
-0.04	2.11E-01	0.326	0.659	0.159	0.238	0.370	0.826	0.960
-0.05	1.43E-01	0.375	0.708	0.208	0.267	0.444	0.875	0.972
-0.06	9.68E-02	0.412	0.745	0.245	0.287	0.504	0.912	0.981
-0.07	6.56E-02	0.438	0.772	0.272	0.302	0.551	0.938	0.987
-0.08	4.44E-02	0.457	0.791	0.291	0.312	0.585	0.957	0.991
-0.09	3.01E-02	0.471	0.804	0.304	0.319	0.610	0.971	0.994
-0.10	2.04E-02	0.480	0.813	0.313	0.323	0.627	0.980	0.996
-0.15	2.91E-03	0.497	0.830	0.330	0.332	0.661	0.997	0.999
-0.20	4.16E-04	0.500	0.833	0.333	0.333	0.666	1.000	1.000

Chapter 6   TRANSIENT METHODS BASED ON POTENTIAL STEPS

Problem 6.6[©] **(a).** Tomeš criterion for a reversible sampled-current voltammogram based on semi-infinite linear diffusion is discussed in Section 6.2.2. It is given as

$$|E_{3/4} - E_{1/4}| = \frac{56.4}{n} \text{mV at } 25°C \tag{1}$$

The shape of a voltammogram for a reversible system in sampled-current voltammetry under semi-infinite linear diffusion conditions is, from equation 6.2.14,

$$E = E_{1/2} + \frac{RT}{nF} \ln \frac{i_{d,c}(\tau) - i(\tau)}{i(\tau)} \tag{6.2.14}$$

at $E_{1/4}$, $i(\tau) = i_{d,c}(\tau)/4$, whereas at $E_{3/4}$, $i(\tau) = 3i_{d,c}(\tau)/4$. Substitution for each into equation 6.2.14 leads to

$$E_{1/4} = E_{1/2} + \frac{RT}{nF} \ln \left[\frac{\frac{3}{4} i_{d,c}(\tau)}{\frac{1}{4} i_{d,c}(\tau)} \right] = E_{1/2} + \frac{RT}{nF} \ln [3] \tag{2}$$

$$E_{3/4} = E_{1/2} + \frac{RT}{nF} \ln \left[\frac{\frac{1}{4} i_{d,c}(\tau)}{\frac{3}{4} i_{d,c}(\tau)} \right] = E_{1/2} - \frac{RT}{nF} \ln [3] \tag{3}$$

Following from Tomeš criterion for a reversible sampled-current voltammogram (equation (1)) at 25 °C,

$$
\begin{aligned}
|E_{3/4} - E_{1/4}| &= \frac{2RT}{nF} \ln [3] \\
&= \frac{2 \times 8.31441 \text{ J mol}^{-1}\text{K}^{-1} \times 298.15 \text{ K}}{n \times 96485 \text{ C mol}^{-1}} \ln [3] \\
&= \frac{0.0564 \text{ V}}{n} = \frac{56.4 \text{ mV}}{n}
\end{aligned}
\tag{4}
$$

(d). To find $|E_{3/4} - E_{1/4}|$ for a totally irreversible, cathodic steady-state voltammogram for a single step, one-electron process, equation 5.4.12 applies for irreversible kinetics. $\Lambda^0 = k^0/m_O$.

$$E = E^{0'} + \frac{RT}{\alpha F} \ln \Lambda^0 + \frac{RT}{\alpha F} \ln \left[\frac{i_{d,c} - i}{i} \right] \tag{5.4.12}$$

At $E_{1/4}$, $i(\tau) = i_d(\tau)/4$, whereas at $E_{3/4}$, $i(\tau) = 3i_d(\tau)/4$. Thus, from equation 5.4.12,

$$E_{1/4} = E^{o'} + \frac{RT}{\alpha F} \ln \Lambda^0 + \frac{RT}{\alpha F} \ln [3] \tag{5}$$

$$E_{3/4} = E^{o'} + \frac{RT}{\alpha F} \ln \Lambda^0 + \frac{RT}{\alpha F} \ln \left[\frac{1}{3} \right] \tag{6}$$

The absolute value of the subtraction of equation (5) from equation (6) leads to

$$|E_{3/4} - E_{1/4}| = \frac{RT}{\alpha F} \ln \frac{1}{9} = \frac{0.564 \, V}{\alpha} = \frac{56.4 \, mV}{\alpha} \tag{7}$$

at 25 °C for $n = 1$. This is in agreement with the wave shape characteristics in Table 5.4.1.

Problem 6.7[©] The keys to this problem are the unit step function, $S_\kappa(t)$, the concept of superposition, and the zero shift theorem. The unit step function multiplies a function $F(t)$, and such that the function is equal to zero until t equals or exceeds κ; for $t \geq \kappa$, the value at time t is defined by $F(t - \kappa)$. Superposition can be used on this double potential step example because (1) the problem can be specified as two separate problems over the ranges of 0 to $t < \kappa$ and $t \geq \kappa$, and (2) the initial and boundary conditions for the second part are known to be independent of the time evolution of the first part. That is, because the surface boundary conditions for the first part are pinned (i.e., either nernstian or zero concentration), the second part can be specified without knowing the time evolution of the surface concentration in the first part. Finally, the zero shift theorem allows the product of the unit step function and $F(t)$ to be transformed into s coordinates.

For $t \leq \tau$, the problem is specified as typical for a potential step to an arbitrary potential E_f, which establishes a value θ'. As the problem involves nernstian surface conditions, it is necessary to specify both O and R. This was specified previously in problem 6.5 by equations 6.1.12 to (10), (6.2.4), and (6.1.9), and in the text by equations (6.5.4) to (6.5.12).

$$\frac{\partial C_O^I(x,t)}{\partial t} = D_O \frac{\partial^2 C_O^I(x,t)}{\partial x^2}$$
$$\frac{\partial C_R^I(x,t)}{\partial t} = D_R \frac{\partial^2 C_R^I(x,t)}{\partial x^2}$$

$$C_O^I(x,0) = C_O^*$$
$$C_R^I(x,0) = C_R^*$$
$$\lim_{x \to \infty} C_O^I(x,t) = C_O^*$$
$$\lim_{x \to \infty} C_R^I(x,t) = C_R^*$$

$$D_O \frac{\partial C_O^I(x,t)}{\partial x}\bigg|_{x=0} = -D_R \frac{\partial C_R^I(x,t)}{\partial x}\bigg|_{x=0} \tag{1}$$

$$\theta' = \frac{C_O^I(0,t)}{C_R^I(0,t)} = \exp\left[\frac{nF}{RT}\left(E_f - E^{0'}\right)\right] \tag{2}$$

$$\frac{i_f(t)}{nFA} = D_O \frac{\partial C_O^I(x,t)}{\partial x}\bigg|_{x=0} \tag{3}$$

As developed in problem 6.5 (equation (18)), this yields the current response for the forward step (equation 6.5.13) where $C_R^* = 0$.

$$\frac{\bar{i}_f(s)}{FA\sqrt{D_O}} = \frac{1}{\sqrt{s}}\left[\frac{C_O^*}{1 + \theta'\xi}\right] \tag{4}$$

Or, on inversion where $\theta' > 0$ and $C_R^* = 0$,

$$\frac{i_f(t)}{nFA\sqrt{D_O}} = \frac{1}{\sqrt{\pi t}}\left[\frac{C_O^*}{1 + \xi\theta'}\right]$$

Also, from problem 6.5 (equation (17)), the concentration of R at the electrode surface on the forward step is found.

$$\overline{C}_R^I(0, s) = \frac{\xi C_O^*}{s\left(1 + \xi\theta'\right)}$$

Or, on inversion,

$$C_R^I(0, t) = \frac{\xi C_O^*}{1 + \xi\theta'} \tag{5}$$

At the electrode surface on the forward step, the concentrations of O and R (C_O' and C_R', respectively) are pinned, such that

$$C_O^I(0, t) = \theta' C_R^I(0, t) = C_O' = \theta' C_R' \tag{6}$$

The surface concentrations for the reverse step are also pinned and thus rigorously specified. These concentrations are independent of the concentration specified on the forward step because the surface condition is nernstian.

$$C_O^{II}(0, t) = \theta'' C_R^{II}(0, t) = C_O'' = \theta'' C_R'' \tag{7}$$

For $t \geq \tau$, define a pair of functions $F_O(x, t)$ and $F_R(x, t)$ that characterize the perturbation in concentration or concentration differential caused by the reverse step.

$$\begin{aligned} F_O(x, t) &= S_\tau(t) C_O^{II}(x, t - \tau) \\ F_R(x, t) &= S_\tau(t) C_R^{II}(x, t - \tau) \end{aligned}$$

Fick's second law applies such that

$$\frac{\partial F_O(x, t)}{\partial t} = D_O \frac{\partial^2 F_O(x, t)}{\partial x^2} \tag{8}$$

$$\frac{\partial F_R(x, t)}{\partial t} = D_R \frac{\partial^2 F_R(x, t)}{\partial x^2} \tag{9}$$

The initial concentrations are both set to zero because there is no perturbation due to the second step at time $t = 0$.

$$F_O(x,0) = 0 \tag{10}$$
$$F_R(x,0) = 0 \tag{11}$$

Also, the perturbation on the reverse step will not significantly affect the bulk concentrations.

$$\lim_{x \to \infty} F_0(x,t) = 0 \tag{12}$$
$$\lim_{x \to \infty} F_R(x,t) = 0 \tag{13}$$

The surface boundary condition is defined by the change in concentration brought about by the reverse step, as well as the temporal shift embedded in $S_\tau(t)$.

$$F_O(0,t) = S_\tau(t) \left[C_O^" - C_O' \right] \tag{14}$$
$$F_R(0,t) = S_\tau(t) \left[C_R^" - C_R' \right] \tag{15}$$

The total flux of O and R at the electrode must be conserved.

$$D_O \frac{\partial C_O(x,t)}{\partial x} \bigg|_{x=0} = -D_R \frac{\partial C_R(x,t)}{\partial x} \bigg|_{x=0} = D_O \frac{\partial C_O^I(x,t)}{\partial x} \bigg|_{x=0} + D_O \frac{\partial F_O(x,t)}{\partial x} \bigg|_{x=0}$$
$$= -D_R \frac{\partial C_R^I(x,t)}{\partial x} \bigg|_{x=0} - D_R \frac{\partial F_R(x,t)}{\partial x} \bigg|_{x=0}$$

Given equation (1),

$$D_O \frac{\partial F_O(x,t)}{\partial x} \bigg|_{x=0} = -D_R \frac{\partial F_R(x,t)}{\partial x} \bigg|_{x=0} \tag{16}$$

The other surface boundary condition is the nernstian condition, specified through equation (7).

By analogy to the solution for the generic semi-infinite case presented in Problem 6.5, substitution of equations (10) through (13) into equations (8) and (9) yields the following expressions in s-coordinates.

$$\overline{F_O}(x,s) = A(s) \exp \left[-\sqrt{\frac{s}{D_O}} x \right] \tag{17}$$
$$\overline{F_R}(x,s) = B(s) \exp \left[-\sqrt{\frac{s}{D_R}} x \right] \tag{18}$$

Chapter 6   TRANSIENT METHODS BASED ON POTENTIAL STEPS

The Laplace transform of equations (14) and (15) and substitution of equations (17) and (18) yields $A(s)$ and $B(s)$.

$$\overline{F_O}(0, s) = \exp\left[-s\tau\right] \frac{C_O'' - C_O'}{s} = A(s)$$

$$\overline{F_R}(0, s) = \exp\left[-s\tau\right] \frac{C_R'' - C_R'}{s} = B(s)$$

Substitution of equations (6) and (7) yields

$$A(s) = \exp\left[-s\tau\right] \frac{\theta'' C_R'' - \theta' C_R'}{s}$$

Application of equation (16) generates the following:

$$-\sqrt{sD_O} \exp\left[-s\tau\right] \frac{\theta'' C_R'' - \theta' C_R'}{s} = \sqrt{sD_R} \exp\left[-s\tau\right] \frac{C_R'' - C_R'}{s}$$

This is solved to find C_R''.

$$C_R'' = C_R' \frac{1 + \xi\theta'}{1 + \xi\theta''}$$

Note that equations (5) and (6) yield the expression for C_R', and thus C_R''.

$$C_R' = \frac{\xi C_O^*}{1 + \xi\theta'}$$

$$C_R'' = \frac{\xi C_O^*}{1 + \xi\theta''}$$

Or,

$$A(s) = \frac{\exp\left[-s\tau\right]}{s}\xi C_O^* \left[\frac{\theta''}{1 + \xi\theta''} - \frac{\theta'}{1 + \xi\theta'}\right]$$

$$B(s) = \frac{\exp\left[-s\tau\right]}{s}\xi C_O^* \left[\frac{1}{1 + \xi\theta''} - \frac{1}{1 + \xi\theta'}\right]$$

Substitution into equations (17) and (18) generates

$$\overline{F_O}(x, s) = \exp\left[-\sqrt{\frac{s}{D_R}}x\right] \frac{\exp\left[-s\tau\right]}{s}\xi C_O^* \left[\frac{\theta''}{1 + \xi\theta''} - \frac{\theta'}{1 + \xi\theta'}\right]$$

$$\overline{F_R}(x, s) = \exp\left[-\sqrt{\frac{s}{D_R}}x\right] \frac{\exp\left[-s\tau\right]}{s}\xi C_O^* \left[\frac{1}{1 + \xi\theta''} - \frac{1}{1 + \xi\theta'}\right]$$

The total concentrations are defined as follows:

$$\overline{C_O}(x, s) = \overline{C_O^I}(x, s) + \overline{F_O}(x, s)$$
$$\overline{C_R}(x, s) = \overline{C_R^I}(x, s) + \overline{F_R}(x, s)$$

The total current on the reverse step is then defined as

$$\frac{\bar{i}_r(s)}{nFA} = -D_R \frac{\partial \overline{C_R}(x, s)}{\partial x}\bigg|_{x=0} = -D_R \frac{\partial \overline{C_R^I}(x, s)}{\partial x}\bigg|_{x=0} - D_R \frac{\partial \overline{F_R}(x, s)}{\partial x}\bigg|_{x=0}$$

From equations (3) and (4),

$$
\begin{aligned}
\frac{\bar{i}_r(s)}{nFA} &= \frac{\bar{i}_f(s)}{nFA} - D_R \frac{\partial \overline{F_R}(x, s)}{\partial x}\bigg|_{x=0} \\
&= \frac{\sqrt{D_O}}{\sqrt{s}}\left[\frac{C_O^*}{1+\theta'\xi}\right] + \sqrt{D_R}\frac{\exp[-s\tau]}{\sqrt{s}}\xi C_O^*\left[\frac{1}{1+\xi\theta"} - \frac{1}{1+\xi\theta'}\right] \\
&= \frac{\sqrt{D_O}}{\sqrt{s}}\left[\frac{C_O^*}{1+\theta'\xi}\right] + C_O^*\sqrt{D_O}\frac{\exp[-s\tau]}{\sqrt{s}}\left[\frac{1}{1+\xi\theta"} - \frac{1}{1+\xi\theta'}\right] \\
&= C_O^*\sqrt{D_O}\frac{\exp[-s\tau]}{\sqrt{s}}\left[\frac{1}{1+\xi\theta"} - \frac{1}{1+\xi\theta'}\right] + \frac{C_O^*\sqrt{D_O}}{\sqrt{s}}\left[\frac{1}{1+\theta'\xi}\right]
\end{aligned}
$$

Or, upon inversion for $t \geq \tau$, equation (6.5.14) is found.

$$\frac{i_r(t)}{nFAC_O^*\sqrt{D_O}} = \frac{1}{\sqrt{\pi(t-\tau)}}\left[\frac{1}{1+\xi\theta"} - \frac{1}{1+\xi\theta'}\right] + \frac{1}{\sqrt{\pi t}}\left[\frac{1}{1+\theta'\xi}\right] \tag{6.5.14}$$

For steps in the forward and reverse directions to the mass transport limited plateaus of reduction of O and oxidation of R, equation (6.5.14) is simplified such that for the reduction $\theta' \to 0$ (equation (2)) and for the oxidation $\theta" \to \infty$ (equation (7)). This yields equation (6.5.15).

$$\frac{i_r(t)}{nFAC_O^*\sqrt{D_O}} = -\frac{1}{\sqrt{\pi(t-\tau)}} + \frac{1}{\sqrt{\pi t}} \tag{6.5.15}$$

Chapter 6   TRANSIENT METHODS BASED ON POTENTIAL STEPS

Problem 6.8[©]   From the data in the caption of Figure 6.6.4, $C_O^* = 1.0 \times 10^{-5}$ mol/cm^3, $A = 0.0230$ cm^2, $t_i^{1/2} = 5.1$ ms$^{1/2}$, and, for a plot of $Q(t)$ vs. $\sqrt{t}$, the slope is 3.52×10^{-6} C/ms$^{1/2}$. The slope is specified by equation (6.6.10).

$$Q(t) = nFAk_fC_O^* \left[\frac{2\sqrt{t}}{H\sqrt{\pi}} - \frac{1}{H^2} \right] \tag{6.6.10}$$

H is expressed in terms of t_i (equation (6.6.11)) as $H = \sqrt{\pi/4t_i}$. (Note that H is also found from the intercept, although with larger error.) Thus, $H = 0.174$ ms$^{-1/2}$. The slope of $Q(t)$ vs. $\sqrt{t}$ is $2nFAk_fC_O^*/(H\sqrt{\pi})$. For Cd^{2+} reduction, $n = 2$.

$$k_f = \frac{3.52 \times 10^{-6} \text{ C/ms}^{1/2} \times 0.174 \text{ ms}^{-1/2} \times \sqrt{\pi} \times 10^3 \text{ ms/s}}{2 \times 2 \times 0.0230 \text{ cm}^2 \times (96485 \text{ C/mol}) \times (1.0 \times 10^{-5} \text{ mol/cm}^3)} = 0.0122 \text{ cm/s}$$

This value agrees well with the value of 0.0116 to 0.0137 cm/s reported in the original paper by Christie, Lauer, and Osteryoung (*JEAC* **7,** 60 (1964)).

=====

Problem 6.10[©]   From the caption of Figure 6.6.3, for the forward step, the slope is 9.89×10^{-6} C/s$^{1/2}$ and the intercept is 7.9×10^{-7} C. From equation (6.6.2) for the forward step, the charge is

$$Q(t) = \frac{2nFAC_O^0\sqrt{D_Ot}}{\sqrt{\pi}} + Q_{dl} + nFA\Gamma_O \tag{6.6.2}$$

From Figure 6.6.1, $n = 1$, $A = 0.018$ cm^2, and $C_O^* = 0.95 \times 10^{-6}$ mol/cm^3. The potential is stepped -260 mV past $E^{0'}$, so that the step is to the mass transport limit. Thus, from $Q(t)$ vs. $\sqrt{t}$,

$$
\begin{aligned}
D_O &= \left[\frac{slope\sqrt{\pi}}{2nFAC_O^*} \right]^2 = \left[\frac{(9.89 \times 10^{-6} \text{ C/}\sqrt{s})\sqrt{\pi}}{2 \times 1 \times (96485 \text{ C/mol})(0.018 \text{ cm}^2)(0.95 \times 10^{-6} \text{ mol/cm}^3)} \right]^2 \\
&= 2.82 \times 10^{-5} \text{ cm}^2\text{/s}
\end{aligned}
$$

Typical values of diffusion coefficients in solution are of the order of 10^{-5} to 10^{-6} cm^2/s, with diffusion coefficients in most volatile (less viscous) organic electrolytes faster than those in water. The most common source of error in calculating diffusion coefficients is using units of M instead of mol/cm^3 for the concentration.

A comparison of equations (6.6.2) and (6.6.6) indicates that the slopes for the forward and reverse steps should be equal if the system is characterized by simple mass transport limited oxidation and reduction. The slope reported for the oxidation is about 5% lower than that for the reduction. The intercepts for the reduction and oxidation are, respectively, 7.9×10^{-7} C and 6.6×10^{-7} C. For a system involving stable, unadsorbed O and R, one expects the slopes and intercepts to be equal. There are differences in this case, but it is not clear whether the differences are statistically significant. The first step in further interpretation should be to determine the experimental precision of the results. Discrepancies at the levels observed here are common with solid electrodes. Some of the contributing factors are discussed in Chapters 14 and 17.

One possible cause of the differences in slopes and intercept is that the oxidized species DCB adsorbs whereas the reduced species DCB$^{\bullet-}$ either does not adsorb or adsorbs less than DCB. If the surface excess for the two forms differ, then this is reflected in the difference in the intercepts for the forward and reverse steps. If the adsorption associated with the forward step is extensive enough, it can disturb the concentration profile sufficiently that the concentration profile of R is disrupted from that expected for a simple mass transport limited reaction. See Section 17.3.1.

An alternative reason for the difference in the slopes is that DCB$^{\bullet-}$ is being consumed through a chemical reaction so that its concentration is less than that of DCB. Here, the formal potential is sufficiently negative that trace oxygen could react with DCB$^{\bullet-}$.

=====

Problem 6.11$^{©}$ **(a).** The time to achieve steady state is dependent on shortest characteristic dimension of the electrode. For a disk or a sphere, the characteristic dimension is the radius r_0. For example, consider equation 6.1.19.

$$i_d\left(t\right) = nFAD_O C_O^* \left[\underbrace{\frac{1}{\sqrt{\pi D_O t}}}_{\text{transient}} + \underbrace{\frac{1}{r_0}}_{\text{steady state}} \right] \tag{6.1.19}$$

For steady state defined as the transient component falling below the steady state component to a fraction represented as $a\%$, equation 6.1.25 results.

$$\frac{a\%}{100} \geq \frac{\sqrt{\pi D_O t}}{r_0} \tag{6.1.25}$$

This rearranged to find the time to steady state.

$$t \geq \left(\frac{a\%}{100}\right)^2 \frac{r_0^2}{\pi D_O}$$

For a large electrode, the time is too long for the system to achieve steady state without thermal disruption of the diffusion layer.

Problem 6.13$^{\copyright}$ (a). The time constant τ_{cell} of 10 ns measured the time scale for capacitive charging of the electrode surface, the nonfaradaic component of the current. $\tau_{cell} = R_{\text{solution}} C_{\text{capacitance}}$ where $i_{\text{charging}}(t)$ decays as $\exp\left[-t/\tau_{cell}\right]$. At $t = \tau_{cell}$, $i_{\text{charging}}(t)$ has decayed to 37% of its initial value; at $t = 3\tau_{cell}$, 5%; at $t = 5\tau_{cell}$, $< 1\%$ (0.7%). At about $5\tau_{call}$ capacitive (nonfaradaic) current is negligible. For τ_{cell} of 10 ns, capacitive charging is diminished to below 1 % at 50 ns. Faradaic measurements can be made after this time.

(b). For $t = 2. \times 10^{-7}$ s and $D = 1. \times 10^{-5}$ cm^{2}s^{-1}, the diffusion length is $\approx \sqrt{2Dt} \approx \sqrt{2\left(1. \times 10^{-5}\ \text{cm}^2\text{s}^{-1}\right)\left(2. \times 10^{-7}\text{s}\right)} = 2. \times 10^{-6}$ cm or 20 nm.

(c). Semi-infinite linear diffusion is a reasonable approximation to the current at a microelectrode provided the diffusion length is $\lesssim 10\%$ of the electrode radius. For r_0 of 1 μm and diffusion length 20 nm, $2. \times 10^{-6}$ cm/$1. \times 10^{-4}$ cm is 0.02 or 2 %, so semi-infinite linear diffusion and the Cottrell equation are appropriate for this experiment. Diffusion to the edge of the electrode and radial diffusion are negligible.

(d). The surface area of a microdisk of 1 μm radius is πr_0^2 or 3×10^{-8} cm^2. The diffusion layer thickness at 200 ns is $2. \times 10^{-6}$ cm. The volume of solution is 3×10^{-8} cm^2 $\left(2. \times 10^{-6}\ \text{cm}\right) \approx 6. \times 10^{-14}$ cm^3. For 1 mM redox species, 1×10^{-6} mol cm^{-3} $\left(6. \times 10^{-14}\ \text{cm}^3\right)\left(6.02 \times 10^{23}\ \text{mol}^{-1}\right) \approx 4 \times 10^4$ molecules of redox probe in the volume.

(e). A monolayer is formed on the electrode surface of 3×10^{-8} cm^2, where the cross sectional area ("parking area" or footprint) of each molecule is $\left(8 \times 10^{-8}\ \text{cm}\right)\left(1.2 \times 10^{-7}\ \text{cm}\right) = 1 \times 10^{-14}$ cm^2. The number of redox molecules on the electrode surface is $\left(3 \times 10^{-8}\ \text{cm}^2\right)\left(1 \times 10^{-14}\ \text{cm}^2\right)^{-1} = 3 \times 10^6$ molecules. The faradaic current at the monolayer modified electrode is likely to be much higher than the current for the 1 mM redox probe diffusing in solution at sufficiently short times.

7 LINEAR SWEEP AND CYCLIC VOLTAMMETRY

Problem 7.1[©] Equations (7.2.8) and (7.2.9) express the surface concentration of O and R in terms of convolution integrals of the current.

$$C_O(0,t) = C_O^* - \frac{1}{nFA\,(\pi D_O)^{1/2}} \int_0^t \frac{i(\tau)}{(t-\tau)^{1/2}} d\tau \tag{7.2.8}$$

$$C_R(0,t) = \frac{1}{nFA\,(\pi D_R)^{1/2}} \int_0^t \frac{i(\tau)}{(t-\tau)^{1/2}} d\tau \tag{7.2.9}$$

These can be combined to yield equation (7.11.1).

$$D_O^{1/2} C_O(0,t) + D_R^{1/2} C_R(0,t)$$

$$= D_O^{1/2}\left[C_O^* - \frac{1}{nFA\,(\pi D_O)^{1/2}} \int_0^t \frac{i(\tau)}{(t-\tau)^{1/2}} d\tau \right] + \frac{D_R^{1/2}}{nFA\,(\pi D_R)^{1/2}} \int_0^t \frac{i(\tau)}{(t-\tau)^{1/2}} d\tau$$

$$= D_O^{1/2} C_O^*$$

=====

Problem 7.3[©] The expression for the peak current in cyclic voltammetry under reversible conditions is given by equation (7.2.20).

$$\frac{i_p(v)}{v^{1/2}} = 0.4463 \left(\frac{F}{RT} \right)^{1/2} F n^{3/2} A D_O^{1/2} C_O^* \tag{7.2.20}$$

For chronoamperometry under mass transport limited conditions, the Cottrell equation (equation (6.1.12)) applies.

$$i(t)t^{1/2} = \frac{nFAC_O^* D_O^{1/2}}{\pi^{1/2}} \tag{6.1.12}$$

Experimental data for cyclic voltammetry and chronoamperometry on a single system will yield both $i_p(v)/v^{1/2}$ and $i(t)t^{1/2}$. The ratio of these two parameters yields an expression for determining n without knowing A, D_O, and C_O^*. In the ratio, A, D_O, and C_O^* cancel.

$$\frac{\frac{i_p^{CV}(v)}{v^{1/2}}}{i(t)^{CA} t^{1/2}} = \frac{0.4463 \left(\frac{F}{RT} \right)^{1/2} F n^{3/2} A D_O^{1/2} C_O^*}{\frac{nFAC_O^* D_O^{1/2}}{\pi^{1/2}}} = 0.4463 \left(\frac{\pi F}{RT} \right)^{1/2} n^{1/2}$$

$$= 4.935 n^{1/2} \text{ V}^{-1/2} \text{ at 298 K}$$

A similar procedure is not suitable for determining n for irreversible reactions. Multistep mechanisms leading to $n > 1$ are intrinsically complex (Section 3.7). There is no way to know how the

Electrochemical Methods: Fundamentals and Applications, Third Edition, Student Solutions Manual. Cynthia G. Zoski and Johna Leddy.
© 2025 John Wiley & Sons Ltd. Published 2025 by John Wiley & Sons Ltd.

Chapter 7   LINEAR SWEEP AND CYCLIC VOLTAMMETRY

CV response depends on n; therefore, one has no basis for deriving $n > 1$ from the ratio as suggested by Mueller and Adams for analysis of reversible systems. If $n = 1$, cyclic voltammetric $i_p / v^{1/2}$ for an irreversible case is given by equation (7.4.5) with $\pi^{1/2} \chi_p (bt) = 0.4958$.

$$\frac{i_p^{CV,\text{irrev}}}{v^{1/2}} = 0.4958 \left(\frac{\alpha F}{RT} \right)^{1/2} FAC_O^* D_O^{1/2} \tag{7.4.5}$$

The Cottrell current in chronoamperometry is independent of the kinetics; therefore, equation (6.1.12) applies with $n = 1$. The diagnostic ratio then becomes

$$\begin{aligned}
\frac{\frac{i_p^{CV,\text{irrev}}(v)}{v^{1/2}}}{i(t)^{CA} t^{1/2}} &= \frac{0.4958 \left(\frac{\alpha F}{RT} \right)^{1/2} FAC_O^* D_O^{1/2}}{\frac{FAC_O^* D_O^{1/2}}{\pi^{1/2}}} = 0.4958 \left(\frac{\pi F}{RT} \right)^{1/2} \alpha^{1/2} \\
&= 5.48 \alpha^{1/2} \text{ V}^{-1/2} \text{ at 298 K}
\end{aligned}$$

Thus, $n = 1$ cannot be confirmed without independent knowledge of α.

It is noted that where n is known equal to 1, the ratio allows determination of α in the irreversible case, without knowledge of A, C_O^*, and D_O.

Where mechanisms are complex, the practical approach is to compare experimental data with computer simulations, as noted in Section 7.4.1(b).

=====

Problem 7.5[©]   Within the potential window of acetonitrile, benzophenone (BP) can only be reduced and TPTA can only be oxidized.

Inspection of cyclic voltammetric peak currents i_{pc} and i_{pa} and the corresponding peak potentials E_{pc} and E_{pa} characterizes homogeneous chemical reactions and heterogeneous electron transfer reactions. Initial inspection for voltammograms with common morphologies derive from $|i_{pa}/i_{pc}|$ and potentials E_{pc} and E_{pa} that characterize the reversibility of the heterogeneous electron transfer (ΔE_p) and the formal potential ($E_{pc} + E_{pa}$).

- For fairly symmetric voltammograms, a peak current ratio $|i_{pa}/i_{pc}|$ of 1 is consistent with no homogeneous chemical reactions (Section 7.2.2(b)). Peak heights are measured from the baseline for the forward and reverse scans, as described in Section 7.2.2(a).

- The peak splitting $\Delta E_p = |E_{pc} - E_{pa}|$ at 25 °C and $n = 1$ identifies the electron transfer as reversible ($\Delta E_p = 59.1/n$ mV, Section 7.2.2(c)), quasireversible ($61 \leq \Delta E_p \leq 212$ mV, from Table 7.3.1 in Section 7.3.2), and irreversible ($\Delta E_p \gtrsim 200$ mV, Section 7.4). The quasireversible range is not rigorously specified; Nicholson considers ΔE_p as large as ≈ 212 mV but Matsuda and Ayabe consider $\Delta E_p \gtrsim 140$ mV. For $\alpha \approx 0.5 \pm 0.2$, quasireversible k^0 is typically between 0.1 and 10^{-5} cm s^{-1}. Note reversibility characterizes the electron transfer rate relative to the mass transport rate, $k^0 (Dfv)^{-1/2}$. For $k^0 (Dfv)^{-1/2}$ large, electron transfer rate is fast compared to mass transport (reversible); for $k^0 (Dfv)^{-1/2}$ small, mass transport outraces electron transfer (irreversible); best measures of heterogeneous electron transfer rate are made where $k^0 (Dfv)^{-1/2}$ is about 1 (quasireversible).

- The formal potential $E^{0\prime}$ is estimated from the half wave potential $E_{1/2}$ as $E^{0\prime} \equiv E_{1/2} - (RT/nF) \ln (D_r/D_O)^{1/2}$, as defined in the footnote on page 315 in Section 7.2.1(a). In most cases, $D_R \approx D_O$ and $E^{0\prime}$ is well approximated by $E_{1/2}$. For reversible electron transfer and no homogeneous reactions, $E_{1/2} = 0.5 (E_{pc} + E_{pa})$ (Section 7.2.2(d)). For quasireversible electron transfer and $\alpha \approx 0.5$, $E_{1/2} \approx 0.5 (E_{pc} + E_{pa})$ (Section 7.3.2(a) and equation (7.3.7)).

(a). For potential scanned from 0.5 to 1.0 V, TPTA is oxidized to the radical cation as TPTA $\rightleftharpoons$ TPTA$^{\bullet+}$ + e. As the potential is scanned in the reverse direction from 1.0 to 0.5 V, the radical cation TPTA$^{\bullet+}$ generated on the forward sweep is reduced back to TPTA.

The voltammogram is fairly symmetric and $|i_{pa}/i_{pc}|$ of 1 is consistent with no homogeneous reactions. The peak splitting $\Delta E_p = |E_{pc} - E_{pa}|$ is about 100 mV, above the approximately $59/n$ mV expected for reversible electron transfers. This peak splitting is consistent with quasireversible electron transfers for scan rates normally accessible at macroscopic electrodes. The potential midpoint between the peak potentials, $0.5\,(E_{pc} + E_{pa})$, is 0.7 V vs QRE. The caption of Figure 7.5.2 notes the QRE is at ≈ -0.03 V vs SCE. The formal potential is 0.67 V vs SCE. The potential axis may illustrate how to convert from QRE to SCE. In Table C.3, the formal potential of TPTA|TPTA$^{\bullet+}$ is 0.98 V vs SCE in THF, which differs from 0.67 V vs SCE. Such differences are not uncommon as redox species are evaluated in different solvents.

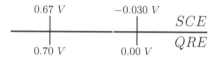

For the potential scanned from –1.5 to –2.0 V, BP is reduced to the radical anion (BP + e $\rightleftharpoons$ BP$^{\bullet-}$) As the potential scan direction is reversed, the radical anion BP$^{\bullet-}$ generated on the forward sweep is oxidized back to BP.

The voltammogram is fairly symmetric and measured peak heights are the same ($|i_{pa}/i_{pc}| \approx 1$), again consistent with no homogeneous chemical reactions. The peak splitting $\Delta E_p \approx 125$ mV, consistent with quasireversible heterogeneous electron transfer at scan rates normally accessible at macroscopic disks. For the same experimental conditions, ΔE_p is larger for BP than TPTA, and k^0 is smaller. Formal potential for BP|BP$^{\bullet-}$ is estimated as. $E^{0\prime} \approx E_{1/2} \approx 0.5\,(E_{pc} + E_{pa})$ is ≈ -1.95 V vs QRE or ≈ -1.92 V vs SCE. The potential axis may illustrate how to convert from QRE to SCE. The $E^{0\prime}$ found for BP|BP$^{\bullet-}$ is in good agreement with formal potential in acetonitrile listed as -1.88 V vs SCE in Table C.

$$
\begin{array}{ccc}
-0.03\ V & -1.92\ V & \\
\underline{\big|\big|} & & SCE \\
\big|\big| & & QRE \\
0.00\ V & -1.95\ V &
\end{array}
$$

(b). The current decays just after i_{pf}, in the potential range from 0.7 to 1.0 V vs QRE. The current in this potential range is decaying because the current is set by the mass transport limited, linear diffusion of the reactant (TPTA) to the planar electrode. That is, the diffusion control of the current for $C_{TPTA} \rightarrow 0$ at the electrode surface causes the current to decay as $t^{-1/2}$. Under mass transport limited, linear diffusion, the flux of material to the electrode decreases with time, as does the current. The same effect is observed in potential step experiments (Chapter 6). In the cyclic voltammogram, this portion of the curve is called the diffusional tail, and as in potential step experiments, the current decays as $t^{-1/2}$.

(c). On the negative going sweep about -0.1 V, the faradaic reduction current for TPTA$^{\bullet-}$ has decayed to zero as -0.1 V is well past E_{pc} for TPTA. On the positive going sweep near -0.1 V, the potential is well positive of E_{pa} for BP$^{\bullet-}$ oxidation. About -0.1 V, faradaic current is negligible. The nonfaradaic current about -0.1 V is charging current. Charging current, i_c, the current to charge the double layer, is set by the double layer capacitance C_d and electrode area A (Section 7.2.1(d)). Equation (7.2.27) $i_c = \pm A C_d v$ can be used to determine electrode area.

From the nonfaradaic capacitive gap of about 1 μA, i_c is about 0.5 μA at $v = 0.1$ V/s. The electrode area is reported as 0.25 cm^2. The capacitance C_d in the acetonitrile electrolyte can be estimated. For A = C/s and V = C/F,

$$C_d = \frac{i_c}{Av} = \frac{0.5\ \mu\text{C s}^{-1}}{\left(0.25\ \text{cm}^{-2}\right)\left(0.1\ \text{V s}^{-1}\right)} = 20\ \mu\text{F cm}^{-2}$$

=====

Problem 7.7[©]   The cyclic voltammogram in Figure 7.11.2 is consistent with the chemically reversible reduction and oxidation of oxygen. The peak splitting of approximately 130 mV is consistent with quasireversible electron transfer kinetics on a cyclic voltammetric timescale for a one electron process. The sampled-transient voltammogram provides a linear plot of E versus $\log\left[(i_d - i)/i\right]$ with a slope of 63 mV, consistent with either a reversible or a highly irreversible electron transfer. (For an irreversible electron transfer under polarographic conditions, $E = E_{1/2} + 0.0542\alpha^{-1/2}\log\left[(i_d - i)/i\right]$; a plot of E versus $\log\left[(i_d - i)/i\right]$ has a slope of $0.0542/\alpha$. See equation (7.2.7) in *Electrochemical Methods*, 2nd edition, 2001.) For reversible electron transfer kinetics on the sampled-transient voltammetric (i.e., polarographic) timescale, equation (1.3.17) applies and the expected slope is RT/nF. As the response on cyclic voltammetric timescale is quasireversible, and the timescale for polarography is longer, the reversible analysis is appropriate for the polarographic data. The ESR signal indicates that the reduction product is a radical.

As small amounts of methanol are added, the voltammogram shifts toward positive potentials and the forward peak increases in height whereas the reverse peak decreases. This behavior is consistent with a chemical reaction between the methanol and the reduction product. The limiting behavior in the presence of methanol shows that the reduction proceeds at -0.4 volts (far positive of that found for oxygen alone). The polarographic current is twice that found in the absence of methanol. The slope of the wave is 78 mV. These results indicate that in the limit, the reduction product is consumed by reaction with methanol. A doubling of the limiting current is consistent with twice as many electrons being transferred in the presence of methanol.

(a). Oxygen is a paramagnetic species with two unpaired electrons. The reduction of oxygen with one electron leads to the formation of superoxide ($O_2^{\cdot -}$), a species with one unpaired electron that is ESR active. The two electron reduction of oxygen leads to peroxide (O_2^{2-}), which is not ESR active. Thus, the reaction being considered is $O_2 + e \rightleftharpoons O_2^{\cdot -}$.

(b). When methanol is present, the current is doubled compared to that in the absence of methanol, but the concentration and other experimental conditions remained the same. The data are consistent with a doubling of n, the number of electrons transferred. The reaction of $O_2^{\cdot -}$ with MeOH will shift the wave to positive potentials. The two electron reduction of oxygen leads to the formation of peroxide, as shown below.

$$O_2 + e \rightleftharpoons O_2^{\cdot -}$$

$$O_2^{\cdot -} + \text{MeOH} \rightleftharpoons \text{HO}_2^{\cdot} + \text{MeO}^-$$

$$\text{HO}_2^{\cdot} + e \rightleftharpoons \text{HO}_2^-$$

$$\text{HO}_2^- + \text{MeOH} \rightleftharpoons \text{H}_2\text{O}_2 + \text{MeO}^-$$

(c). The cyclic voltammogram shown in Figure 7.11.2 is consistent with quasireversible electron transfer kinetics. The sampled transient voltammetric data are consistent with an almost reversible electron transfer as shown by the slope of 63 mV. The measurement timescale for sampled transient voltammetry is longer than for cyclic voltammetry. The timescale for the measurement by cyclic voltammetry is quasireversible as the rates of electron transfer and mass transport are comparable. With the longer timescale of sampled transient voltammetry, the measurement is slow compared to the electron transfer rate and rate is classified as reversible (fast) for sampled transient voltammetry.

(d). In summary, oxygen alone undergoes a one electron reduction to superoxide radical anion. On a cyclic voltammetric timescale, this reduction is quasireversible; on a sampled transient voltammetric (polarographic) timescale, the reduction is reversible. In the presence of methanol, the oxygen undergoes a two electron reduction to form peroxide. The reaction of the reduction product with methanol shifts the reduction to less extreme potentials. The polarographic data suggest that the kinetics for this process are less than reversible.

The difference in oxygen reduction in the absence and presence of protons is highlighted in this problem. In water, oxygen is reduced by two or four electrons to either peroxide or water. Superoxide is not generated.

=====

Problem 7.9[©] For 1,3,5-tri-*tert*-butylpentalene (ttBP) denoted as **I,** polarographic data are consistent with reversible electrolysis on reduction of **I** (R.W. Johnson JACS **99** 1461 (1968)). From the cyclic voltammogram in Figure 7.11.4, peak current for **I** oxidation is at ≈ 0.75 V and reduction is at ≈ -1.5 V. Both waves are consistent with no homogeneous reactions.

(a). **I** can be oxidized to form $\mathbf{I^{+\cdot}}$, a green radical cation, as indicated by the ESR signal.

$$\mathbf{I} \rightleftharpoons \mathbf{I^{+\bullet}} + e \qquad\qquad E_{1/2} \approx +0.8 \text{ V vs. SCE}$$

This can be reduced back to **I**. From the cyclic voltammogram, **I** can be reduced to form a magenta solution of radical anion $\mathbf{I^{-\cdot}}$,

$$\mathbf{I} + e \rightleftharpoons \mathbf{I^{-\bullet}} \qquad\qquad E_{1/2} \approx -1.46 \text{ V}$$

This is consistent with an ESR signal. The radical anion can be oxidized to regenerate **I**.

(b). The cyclic voltammetric wave at approximately $+ 0.75$ V is consistent with the reaction ($\mathbf{I} \rightleftharpoons \mathbf{I^{+\cdot}} + e$). The electron transfer is a single electron, reversible process as indicated by a peak splitting of approximately 60 mV. There is no evidence of homogeneous reactions. The wave at approximately -1.5 volts is also chemically reversible. The peak splitting is approximately 190 mV for the reduction wave, which suggests quasireversible/irreversible nature to this reaction. This separation suggests that the kinetic parameter ψ (equation (7.3.6)) is, from Table 7.3.1, approximately 0.10 to 0.20 or Λ is approximately 0.2 to 0.4.

(c). For the reduction, the cyclic voltammogram has a peak splitting consistent with quasireversible (approaching irreversible) kinetics. The slope of the wave of the polarographic experiment is 59 mV, consistent with either highly irreversible or reversible electron transfer. (For an irreversible electron transfer under polarographic conditions, $E = E_{1/2} + 0.0542\alpha^{-1/2} \log\left[(i_d - i)/i\right]$. See equation (7.2.7) in *Electrochemical Methods*, 2nd edition, 2001. A slope of 59 mV yields an improbable α of 0.9, which favors a reversible electron transfer reaction. Cyclic voltammograms would be highly asymmetric for α of 0.9.) The wave shown in Figure 7.11.4 is symmetric, consistent with a transfer coefficient of 0.5 ± 0.2, or 0.3. The polarographic data are consistent with reversible electron transfer. The quasireversible behavior observed under cyclic voltammetric perturbation is consistent with the much faster timescale associated with a measurement at 500 mV/s. The term reversibility characterizes a system where the rate of heterogeneous electron transfer is rapid compared to the rate of voltammetric perturbation. Chemically reversible refers to a process where there are no chemical reactions that perturb the concentration of the electroactive species on the timescale of the measurement.

(d). For the reduction, the peak splitting at 500 mV/s is about 190 mV. From Table 7.3.1, this yields ψ of approximately 0.14. For $D_O = D_R = 2 \times 10^{-5}$ cm^2/s, equation (7.3.6) simplifies to

$$k^0 = \psi \left(\frac{\pi D_O F v}{RT} \right)^{1/2} = \psi \sqrt{122.3 D_O v} \text{ at 298 K}$$

Here, $k^0 = 0.005$ cm/s. For different scan rates, this can be used to calculate ψ and then find ΔE_p from Table 7.3.1.

Working curves presented by Matsuda and Ayabe, *Z. Elektrochem.*, **59** 494 (1955), are shown in *Electrochemical Methods*, 2nd edition (2001) in Figure 6.4.2 for $K(\Lambda, \alpha)$ versus $\log \Lambda$; $\Lambda = \psi \pi^{1/2}$. From the plot, $\log \Lambda$ yields $K(\Lambda, \alpha)$ where $i_p = i_p(rev) K(\Lambda, \alpha)$. Equation (7.2.21) specifies $i_p(rev)$. The plots of i_p and ΔE_p with v evolve from the working curves.

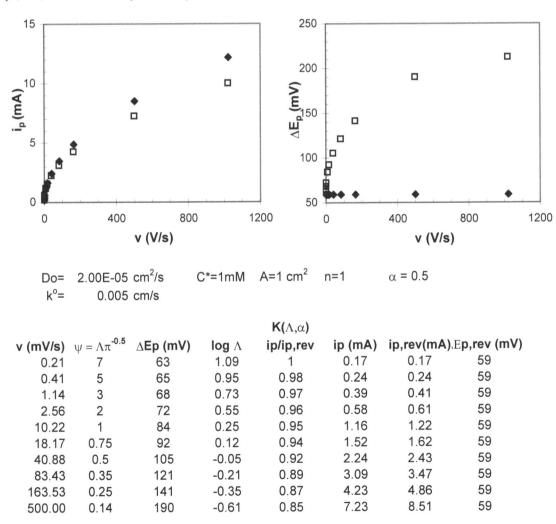

v (mV/s)	$\psi = \Lambda\pi^{-0.5}$	ΔEp (mV)	$\log \Lambda$	$K(\Lambda,\alpha)$ ip/ip,rev	ip (mA)	ip,rev(mA)	Ep,rev (mV)
0.21	7	63	1.09	1	0.17	0.17	59
0.41	5	65	0.95	0.98	0.24	0.24	59
1.14	3	68	0.73	0.97	0.39	0.41	59
2.56	2	72	0.55	0.96	0.58	0.61	59
10.22	1	84	0.25	0.95	1.16	1.22	59
18.17	0.75	92	0.12	0.94	1.52	1.62	59
40.88	0.5	105	-0.05	0.92	2.24	2.43	59
83.43	0.35	121	-0.21	0.89	3.09	3.47	59
163.53	0.25	141	-0.35	0.87	4.23	4.86	59
500.00	0.14	190	-0.61	0.85	7.23	8.51	59

The peak splitting is pinned at 59 mV. Allow $\alpha = 0.5$; $C_O^* = 1$ mM; $A = 1$ cm^2; $n = 1$. Plots of i_p and ΔE_p with v are shown for both the oxidation and reduction of **I**. The reversible oxidation process is marked as closed diamonds; it is assumed that the process remains reversible at 1 V/s. The reduction is marked with open squares. Note that over the range of scan rates reported below, the reduction response varies from almost reversible to irreversible. The development below is more precise than the sketches requested in the problem.

To summarize, for the reversible green couple, $i_{pa}(\mathbf{I}, \mathbf{I^{+\bullet}})$ varies as $v^{1/2}$ whereas $\Delta E_p(\mathbf{I}, \mathbf{I^{+\bullet}})$ is constant at approximately 59 mV. For the quasireversible magenta couple, $i_{pc}(\mathbf{I}, \mathbf{I^{-\bullet}})$ variation with v is described by Matsuda and Ayabe as shown. ΔE_p will increase with increasing scan rate as given in Table 7.3.1.

=====

Problem 7.11© **(a).** The value of ψ is calculated using equation (7.3.6). When $D_O = D_R$, this reduces to

$$\psi = \frac{k^0}{\sqrt{\pi D_O F v / RT}}$$

Mirkin, Richards, and Bard found $k^0 = 3.7$ cm/s and $D_R = 1.70 \times 10^{-5}$ cm^2/s. Paul and Leddy (*Anal. Chem.* **67**(10) (1995) 1661-1668) have reported a simple linear relationship between ψ and ΔE_p. The relationship varies slightly with α, but is well generalized for $0.5 \leq \alpha \leq 0.7$ as follows. It is formally applicable for $0.1 \leq \psi \leq 20$. The transfer coefficient for ferrocene is close to 0.5.

$$\frac{nF}{RT}\Delta E_p = \frac{0.779}{\psi} + 2.386$$

Or, for 298 K and $n = 1$,

$$\Delta E_p = \frac{0.0200}{\psi} + 0.06131$$

Values of ψ and ΔE_p are tabulated below. For comparison, ΔE_p found by extrapolation from the data of Nicholson and Shain in Table 7.3.1 are listed.

v (V/s)	ψ	ΔE_p (mV)	ΔE_p (mV) (NS)
3	46.85	61.7	
30	14.82	62.7	61.8
100	8.12	63.8	62.8
200	5.74	64.8	64.3
300	4.69	65.6	65.3
600	3.31	67.3	67.4

(b). The values calculated in part (a) are based on linear diffusion to a planar electrode. The first question is whether the 12.5 μm radius electrode used by Noviandri, et. al. satisfies the requirements for linear diffusion over the range of scan rates from 3.2 to 640 V/s. The diffusion length is set by $[RTD_O/(nFv)]^{1/2}$. Here, the diffusion length for the slowest scan rate is 3.7 μm. As the diffusion length is less than the radius, the scan rates are fast enough to allow the assumption of linear diffusion used above.

Chapter 7   LINEAR SWEEP AND CYCLIC VOLTAMMETRY

Solution resistance (R) can distort cyclic voltammetric waves, and increase the peak splitting. Crudely, the increase in peak splitting is approximated by Ohm's law as $2|i_p(v)|R$. The electrolyte of 0.1 M TBABF$_4$ in acetonitrile typically has a solution resistance of a few ohms. The peak currents for these electrodes do not exceed a few microamps. (The currents reported in Figure 4 of the original reference, as calculated from equation (7.2.21), would appear to be consistent with a radius of 25 μm as opposed to the reported diameter of 25 μm.) For a few ohms of resistance, the increase in peak splitting would be less than a millivolt at the highest scan rate. The peak splitting reported in part (b) and the values calculated in part (a) differ by 15 to 233 mV. For currents of a few microamps, solution resistance in excess of 10 kΩ would be anticipated. Noviandri, et. al. note that no attempt was made to compensate for solution resistance.

Alternatively, the data provided in part (b) can be analyzed to determine the corresponding standard heterogeneous electron transfer rate. Reversing the process outlined in part (a), the following rate constants are determined.

v (V/s)	ΔE_p (mV)	ψ	k^0 (cm/s)
3.2	77	1.275	0.104
32	94	0.612	0.158
102	96	0.577	0.265
204	120	0.341	0.222
297	134	0.275	0.216
320	158	0.207	0.169
640	300	0.084	0.097

The standard heterogeneous rate constant falls with scan rate as scan rates exceed 100 V/s, which is consistent with uncompensated resistance. That the standard rate is lower at lower scan rates is not consistent with a resistance effect. The behavior at lower scan rates could arise if there were a following chemical reaction that is out-raced at faster scan rates.

Ferrocene is usually viewed as an ideal redox couple, and it exhibits almost ideal voltammetric responses at scan rates traditionally used at macroscopic electrodes ($v \lesssim 10$ V/s). Faster scan rates accessible at microelectrodes allow the determination of heterogeneous electron transfer rates, but the faster scan rates also assay fast chemical reaction kinetics, including adsorption. The adsorption/desorption step would be included in the measured rate of heterogeneous electron transfer, and would tend to depress the measured rate from its true value. Higher electrolyte concentration or a change in electrode material may be sufficient to suppress the adsorption. Electron transfer can be suppressed if the electrode material itself is even slightly oxidized or reduced over the potential range of cyclic voltammetric perturbation.

From the information available, it is not entirely clear why the rates measured by Noviandri, et. al., are about twentyfold lower than those measured by Mirkin, et. al. However, it is the usual assumption that the slower measured rate is more likely to be compromised by an experimental complexity such as resistance. The idea is that there are many ways to slow a measured rate, but only a few to enhance it. The ways in which the rate constant can be incorrectly determined using microelectrodes are usually associated with a poor or incomplete appreciation of the electrode area and geometry. The smallest electrodes, prized for the enhanced transport rates that allow the measurement of very fast heterogeneous rates, are also the most difficult to characterize geometrically.

Problem 7.12[C] The proposed mechanism combines an electron transfer for TEMPO$^\bullet$|TEMPO$^+$ with complexation between TEMPO$^+$ and azide, N$_3^-$.

$$TEMPO^+ + e \;\rightleftharpoons\; TEMPO^\bullet \qquad E_{1/2} = 0.25 \text{ V vs Fc}^+|\text{Fc} \qquad (7.2.43)$$

$$TEMPO^+ + N_3^- \;\rightleftharpoons\; \left(TEMPO^+ N_3^-\right) \quad K \qquad\qquad\qquad (7.2.44)$$

The reaction is written as the reduction, but the starting material is the radical TEMPO$^\bullet$. The TBAN$_3$ is ionized on dissolution in acetonitrile containing 0.1 M LiClO$_4$. As azide increases from 1 to 10 mM, $E_{1/2} = 0.5\left(E_{pc} + E_{pa}\right)$ shifts -60 mV toward the starting potential of -0.2 V. Also, $\Delta E_p = E_{pa} - E_{pc}$ narrows slightly as the electron transfer rate increases at $v = 0.1$ V s^{-1}. Across all concentrations, i_{pa}, i_{pc}, and $|i_{pa}/i_{pc}|$ of 1 are invariant with concentration. The electrode is a 1.5 mm glassy carbon disk. Plot of $E_{1/2}$ against $\log(C_{azide}/\text{mM})$ has slope -62 mV and intercept at 1 mM of 210 mV.

(a) For O, R, A, and OA denoting $TEMPO^+$, $TEMPO^\bullet$, N_3^-, and $\left(TEMPO^+ N_3^-\right)$, the equilibrium reactions are re-expressed. $n = 1$.

$$O + e \;\rightleftharpoons\; R \qquad E_{1/2} \qquad\qquad (1)$$

$$O + A \;\rightleftharpoons\; OA \qquad K \qquad\qquad (2)$$

From Section 2.1.7, the equilibrium potential E_{eq} is expressed by the standard potential E^0 and activities a_O and a_R of the oxidized ($TEMPO^+$) and reduced ($TEMPO^\bullet$) forms of TEMPO. E_{eq} is equivalently characterized by the formal potential $E^{0\prime}$ and corresponding concentrations for O and R. For equation (1), E_{eq} is defined.

$$E_{eq} = E^{0\prime} + \frac{RT}{F} \ln \frac{[O]}{[R]} = E^0 + \frac{RT}{F} \ln \frac{a_O}{a_R} \qquad (3)$$

Let the molar concentration of free (uncomplexed) O and R be c_O and c_R. Denote activity coefficients for O and R of γ_O and γ_R. For a standard state concentration C^0, activities are defined as $a_O = \gamma_O c_O/C^0$ and $a_R = \gamma_R c_R/C^0$. Define $C^0 = 1$ M.

The C^0 normalization yields unitless analytical concentrations $[O]$ and $[R]$. For R, only one form (R) is present. For O, two forms are present, O and OA. Denote the molar concentration of the complex OA as c_{OA}.

$$[R] \;=\; \frac{c_R}{C^0}$$

$$[O] \;=\; \frac{c_O + c_{OA}}{C^0}$$

Let the molar concentration of free (uncomplexed) N_3^- be c_A. Then, $a_A = \gamma_A c_A/C^0$. a_A, γ_A, and $[A]$ are the activity, activity coefficient, and concentration of uncomplexed N_3^-. Let the unitless total analytical concentration of azide be $\left[N_3^-\right]$.

$$\left[N_3^-\right] = \frac{c_A + c_{OA}}{C^0} \qquad (4)$$

The thermodynamic equilibrium constant is defined by activities. From equation (2), K is expressed. Define $a_{OA} = \gamma_{OA} c_{OA}/C^0$.

$$K = \frac{a_{OA}}{a_O a_A} = \frac{\gamma_{OA} c_{OA}/C^0}{\left(\gamma_O c_O/C^0\right)\left(\gamma_A c_A/C^0\right)} = C^0 \frac{\gamma_{OA}}{\gamma_O \gamma_A} \frac{c_{OA}}{c_O c_A} \qquad (5)$$

Solve equation (3) for the formal potential. Substitute $a_O = \gamma_O c_O / C^0$ and $a_R = \gamma_R c_R / C^0$ to find $E^{0\prime}$ in terms of standard potential E^0, activity coefficients γ_O and γ_R, and concentrations c_O, c_R, and c_R.

$$
\begin{aligned}
E^{0\prime} &= E^0 + \frac{RT}{F} \ln \frac{a_O}{a_R} + \frac{RT}{F} \ln \frac{[R]}{[O]} \\
&= E^0 + \frac{RT}{F} \ln \frac{\gamma_O c_O / C^0}{\gamma_R c_R / C^0} + \frac{RT}{F} \ln \frac{[R]}{[O]} \\
&= E^0 + \frac{RT}{F} \ln \frac{\gamma_O}{\gamma_R} + \frac{RT}{F} \ln \frac{c_O [R]}{c_R [O]} \\
&= E^0 + \frac{RT}{F} \ln \frac{\gamma_O}{\gamma_R} + \frac{RT}{F} \ln \frac{c_O \frac{c_R}{C^0}}{c_R \frac{c_O + c_{OA}}{C^0}} \\
&= E^0 + \frac{RT}{F} \ln \frac{\gamma_O}{\gamma_R} + \frac{RT}{F} \ln \frac{c_O}{c_O + c_{OA}} \\
E^{0\prime} &= E^0 + \frac{RT}{F} \ln \frac{\gamma_O}{\gamma_R} - \frac{RT}{F} \ln \left(1 + \frac{c_{OA}}{c_O} \right)
\end{aligned}
$$

Solve equation (5) for c_{OA}/c_O.

$$
\frac{c_{OA}}{c_O} = K \frac{c_A}{C^0} \frac{\gamma_O \gamma_A}{\gamma_{OA}} \tag{6}
$$

Rearrange equation (4) to solve for $c_A / C^0 = [N_3^-] - c_{OA}/C^0$. Substitute to find c_{OA}/c_O in terms of $[N_3^-]$.

$$
\frac{c_{OA}}{c_O} = K \frac{\gamma_O \gamma_A}{\gamma_{OA}} \left([N_3^-] - \frac{c_{OA}}{C^0} \right)
$$

Substitute this into the expression for $E^{0\prime}$.

$$
E^{0\prime} = \underbrace{E^0 + \frac{RT}{F} \ln \frac{\gamma_O}{\gamma_R}}_{E_0^{0\prime}} - \frac{RT}{F} \ln \left(1 + K \frac{\gamma_O \gamma_A}{\gamma_{OA}} \left([N_3^-] - \frac{c_{OA}}{C^0} \right) \right) \tag{7}
$$

Equation (7) is the general equation for the reaction sequence (1) and (2).

(b) The limiting case where no azide is present, $[N_3^-] = 0$, tests the derivation. For analytic concentration, $[N_3^-] = 0$, no complex is formed and $c_{OA} = 0$.

$$
\begin{aligned}
E^{0\prime} &= E^0 + \frac{RT}{F} \ln \frac{\gamma_O}{\gamma_R} - \frac{RT}{F} \ln (1) \\
E^{0\prime} &= E^0 + \frac{RT}{F} \ln \frac{\gamma_O}{\gamma_R}
\end{aligned}
$$

This is the definition of the formal potential, equation (2.1.46).

(c) Under ideal conditions, activity coefficients are equal to 1. For conditions where the fraction of complex is small, $[N_3^-] \gg \frac{c_{OA}}{C^0}$, Equation (7) is simplified.

$$
E^{0\prime} = E_0^{0\prime} - \frac{RT}{F} \ln \left(1 + K [N_3^-] \right)
$$

$E_0^{0\prime}$ is the formal potential in the azide free solution.

(**d**) A limiting form at unit activity coefficients is found where $K\left[N_3^-\right] >> 1$.

$$E^{0\prime} = E_0^{0\prime} - \frac{RT}{F} \ln K\left[N_3^-\right] \tag{8}$$

If $K\left[N_3^-\right]$ is comparable to 1, $E^{0\prime}$ is proportional to $\ln\left(1 + K\left[N_3^-\right]\right)$. If $K\left[N_3^-\right] << 1$, $E^{0\prime} \approx E_0^{0\prime}$.

(**e**) The ordinate (y axis) in Figure 7.2.6b is $0.5\left(E_{pc} + E_{pa}\right)$. The formal potential is estimated from the half wave potential $E_{1/2}$ as $E_{1/2} - (RT/nF)\ln\left(D_R/D_O\right)^{1/2}$, as defined in the footnote on page 315 in Section 7.2.1(a). $E_{1/2} = 0.5\left(E_{pc} + E_{pa}\right)$.

Here, $n = 1$ and the formal potential is for no azide present, $E_0^{0\prime} \equiv E_{1/2} - (RT/F)\ln\left(D_R/D_O\right)^{1/2} = 0.5\left(E_{pc} + E_{pa}\right) - (RT/F)\ln\left(D_R/D_O\right)^{1/2}$. Add $(RT/F)\ln\left(D_R/D_O\right)^{1/2}$ to both sides of Equation (8).

$$E^{0\prime} + (RT/F)\ln\left(D_R/D_O\right)^{1/2} = E_0^{0\prime} + (RT/F)\ln\left(D_R/D_O\right)^{1/2} - \frac{RT}{F}\ln K\left[N_3^-\right]$$

$$E_{1/2} = E_0^{0\prime} + (RT/F)\ln\left(D_R/D_O\right)^{1/2} - \frac{RT}{F}\ln K\left[N_3^-\right]$$

$$0.5\left(E_{pc} + E_{pa}\right) = E_0^{0\prime} + (RT/F)\ln\left(D_R/D_O\right)^{1/2} - \frac{RT}{F}\ln K\left[N_3^-\right]$$

From equation (7.2.3), $E_{1/2}$ is 0.25 V vs Fc$^+$/Fc, with no azide present. . Denote this half wave potential with no azide present as $E_{1/2}(0)$. If $D_R \approx D_O$, then $E_{1/2}(0) \approx E_0^{0\prime} = +0.25$ V.

(**f**) The linearized expression $y = mx + b$ for the data in Figure 7.2.6b is defined for a plot of $0.5\left(E_{pc} + E_{pa}\right)$ versus $\log\left[N_3^-\right]$.

$$\underbrace{0.5\left(E_{pc} + E_{pa}\right)}_{y} = \underbrace{E_0^{0\prime} + (2.303RT/nF)\log\left(D_r/D_O\right)^{1/2} - \frac{2.303RT}{F}\log K}_{\text{intercept}} \underbrace{- \frac{2.303RT}{F}}_{\text{slope}}\underbrace{\log\left[N_3^-\right]}_{x}$$

The plot is linear. The experimental slope is reported as -62 mV that is consistent with $-2.303RT/F = -59$ mV. The experimental results are consistent with the model in equations 7.4.23 and 7.2.44.

(**g**) From equation (9), the slope is $-2.303RT/F$ or 59 mV at 25 °C and for $D_R \approx D_O$, the intercept is $E_0^{0\prime} - \frac{2.303RT}{F}\log K$.

From the plot in Figure 7.2.6b, and the slope is -62 mV. The intercept is reported at $\left[N_3^-\right]$ of 1×10^{-3} is 210 mV. For $\left[N_3^-\right] = 1$, the intercept is lowered by three decades or $62 \times 3 = 186$ mV. The intercept will be $(210 - 186)$ mV $= 24$ mV vs Fc$^+$|Fc at $\left[N_3^-\right] = 1$.

Equation 7.2.43 reports $E_{1/2} = 250$ mV vs Fc$^+$|Fc, where $E_{1/2}$ is the halfwave potential for TEMPO with no azide present. $E_{1/2}$ is $E_0^{0\prime}$. Given $E_0^{0\prime} = +250$ mV. At $\left[N_3^-\right] = 1$, $\log\left[N_3^-\right] = 0$. From equation (9) for $D_R \approx D_O$, $E_0^{0\prime} = +250$ mV, and $\left[N_3^-\right] = 1$,

$$\frac{E_{pc} + E_{pa}}{2} = 24 = 250 - 59\log K$$

$$\log K = \frac{24 - 250}{-59} = 3.8$$

$$K = 10^{3.8} \approx 7 \times 10^3$$

(h) In part (c), the assumption is made that $\left[N_3^-\right] \gg c_{OA}/C^0$. To test its validity, one must examine the probable concentrations of the key participants near the electrode, as determined by the equilibrium.

$$O + A \rightleftharpoons OA$$

The equilibrium constant is

$$K = \frac{c_{OA}/C^0}{(c_O/C^0)\,(c_A/C^0)} \tag{10}$$

At any location in solution

$$
\begin{aligned}
[O] &= \frac{c_O + c_{OA}}{C^0} \\
\left[N_3^-\right] &= \frac{c_A + c_{OA}}{C^0}
\end{aligned}
$$

Therefore, equation (10) can be written as

$$K = \frac{c_{OA}/C^0}{\left([O] - c_{OA}/C^0\right)\left(\left[N_3^-\right] - c_{OA}/C^0\right)} \tag{11}$$

If $\left[N_3^-\right]$ and $[O]$ are known or can be reasonably approximated, then equation (11) is an equation in one unknown that can be solved to find c_{OA}. In fact, $\left[N_3^-\right]$ is precisely known as the added azide concentration. The value of $[O]$ varies with spatial position and electrode potential, but near an electrode controlled at $E_{1/2}$, it should be about 0.5×10^{-3} for the systems examined in the study of Figure 7.2.6, 1 mM TEMPO$^\bullet$ in the bulk. Therefore,

$$7000 = \frac{c_{OA}/1\,\text{M}}{\left(5 \times 10^{-4} - c_{OA}/1\,\text{M}\right)\left(\left[N_3^-\right] - c_{OA}/1\,\text{M}\right)} \tag{12}$$

Given $\left[N_3^-\right]$, one can find the value of c_{OA} that satisfies equation (12). A simple way to reach a solution is by successive approximation using a spreadsheet. One simply determines the value of c_{OA} that allows the right-hand side of (11) to equal 7000.

The results are summarized in the table below.

$\left[N_3^-\right]$	c_{OA} (M)	Computed K	$\left[N_3^-\right] C^0/c_{OA}$
1.00×10^{-3}	4.04×10^{-4}	7010	2.5
2.00×10^{-3}	4.58×10^{-4}	7000	4.4
3.00×10^{-3}	4.73×10^{-4}	6990	6.3
5.00×10^{-3}	4.85×10^{-4}	7020	10.3
10.00×10^{-3}	4.93×10^{-4}	7000	20.3

For $\left[N_3^-\right] \geq 3 \times 10^{-3}$, the ratios in the last column exceed 6 and reach more than 20, so the assumption that $\left[N_3^-\right] \gg c_{OA}C^0$ is reasonably valid over most of the experimental range. Toward lower concentrations of azide, it becomes progressively less so.

In part (d), the assumption is made that $K\left[N_3^-\right] \gg 1$. For $K = 7000$, $K\left[N_3^-\right]$ is 7 for the lowest added azide concentration and exceeds an order of magnitude for all higher concentrations in the table. The assumption is sufficiently valid for the analysis carried out here.

8 POLAROGRAPHY, PULSE VOLTAMMETRY, AND SQUARE-WAVE VOLTAMMETRY

Problem 8.1[©] The wave shape for sampled current voltammetry as described by equation (6.2.14) applies also to normal pulse voltammetry (NPV) for a reversible system, as is the case in this problem. Equation (6.2.14) can be written

$$E = E_{1/2} + \frac{RT}{nF} \ln \frac{i_{d,c}(\tau) - i(\tau)}{i(\tau)} \tag{6.2.14}$$

Thus, a plot of E versus $\ln \frac{i_{d,c}(\tau) - i(\tau)}{i(\tau)}$ should be linear with a slope of $0.02569/n$ V at 25 °C and have an intercept of $E_{1/2}$. From equation (6.2.13),

$$E_{1/2} = E^{0\prime} + RT/(nF) \ln \sqrt{\frac{D_R}{D_O}} \tag{6.2.13}$$

For $D_O = D_R$, $E_{1/2} = E^{0\prime}$. Linear regression of the data provides a slope and an intercept.

(a). $Slope = 0.01265 = \frac{0.02569}{2}$ V, so that $n = 2.03 = 2$ electrons.

(b). $Intercept = E_{1/2} = -0.417$ V vs SCE. As $D_O = D_R$, the formal potential, $E^{0\prime} = E_{1/2}$.

=====

Problem 8.3[©] The reaction under consideration is A + ne $\rightleftharpoons$ B, where $E^A_{1/2} = -1.90$ V vs SCE, the wave slope = 60.5 mV, and $(I)_{max} = 2.15$. When C (dibenzo-15-crown-5) is added to the solution, the wave slope does not change significantly. From Section 5.3.1 (a), the wave slope refers to the slope from a plot of E vs $\log[(i_{d,c} - i)/i]$, which, for a reversible system, should be $59.1/n$ mV at 25 °C.

(a). The value of $(I)_{max} = 2.15$ lies in the usual range for a $1e$ process. This is a strong indication that $n = 1$; therefore, the wave slope near 60 mV indicates that the electrode reaction is reversible.

(b). Equation (8.1.6) applies.

$$(I)_{max} = 708nD_A^{1/2} = 2.15 = 708D_A^{1/2} \tag{8.1.6}$$

Solving for D_A leads to $D_A = (2.15/708)^2 = 9.2 \times 10^{-6}$ cm^2/s, which is a reasonable value for 1e reactions and species in solution.

(c). Following Section 5.3.2 (c) for reversible i-E curves, the shift of $E_{1/2}$ with concentration of C suggests complexation of A with C. The fact that $(I)_{max}$ remains almost constant indicates that n remains at a value of 1.

$$A + pC \rightleftharpoons AC_p \qquad\qquad K_c = \frac{C_{AC_p}}{C_A C_C^p}$$

$$A + e \rightleftharpoons B \qquad\qquad E^{0\prime,A}$$

Electrochemical Methods: Fundamentals and Applications, Third Edition, Student Solutions Manual. Cynthia G. Zoski and Johna Leddy. © 2025 John Wiley & Sons Ltd. Published 2025 by John Wiley & Sons Ltd.

Equation (5.3.11) at 25 °C is expressed without and with ligands as A and AC_p. Note K_c is a formation constant but K in equation (5.3.9) is a dissociation constant; $K = K_c^{-1}$.

$$E_{1/2}^{AC_p} = E^{0',A} + \frac{0.0591}{n} \log \frac{m_B}{m_{AC_p}} - \frac{0.0591}{n} \log K_c - \frac{0.0591p}{n} \log C_C^* \qquad (1)$$

m_B and m_{AC_p} are transport coefficients. On substitution of $E_{1/2}^A = E^{0',A} + \frac{0.0591}{n} \log \frac{m_B}{m_A}$ (equation (1.3.16)), equation (1) is reexpressed in values of $E_{1/2}$.

$$E_{1/2}^{AC_p} = E_{1/2}^A + \frac{0.0591}{n} \log \frac{m_A}{m_{AC_p}} - \frac{0.0591}{n} \log K_c - \frac{0.0591p}{n} \log C_C^* \qquad (2)$$

Equation (8.1.3) describes $(i_d)_{\max}$, the maximum current measured at $t_{\max}$, just before the drop falls. From equation (8.1.6), $(i_d)_{\max}$ is related to $(I)_{\max}$. The mass transport coefficient is proportional to the maximum current at the mass transport limit (equation (1.3.10)). By analogy, the mass transport coefficient for A, m_A is then expressed such that $m_A \propto (I)_{\max}$.

$$m_A = 708 D_O^{1/2} m^{2/3} t_{\max}^{1/6} = (I)_{\max} m^{2/3} t^{1/6} \qquad (3)$$

For m and t constant towards the end of the drop life just before the drop falls, the ratio of the mass transport coefficients for A and AC_p) is found.

$$\frac{m_A}{m_{AC_p}} = \frac{(I)_{\max,A}}{(I)_{\max,AC_p}} = \frac{2.15}{(I)_{\max}} \qquad (4)$$

$(I)_{\max,AC_p}$ is tabulated as $(I)_{\max}$ in the problem for each ligand concentration of C_C^*. With no ligand, $E_{1/2}^A = -1.90$ V vs SCE, and $n = 1$. Substitution into equation ((2)) of $E_{1/2}^{AC_p}$, $E_{1/2}^A$, C_C^*, and m_A/m_{AC_p} from equation ((4)) allows determination of p and then K_c.

Linear regression analysis of the data $E_{1/2}^{AC_p} - E_{1/2}^A$ vs $\log C_C^*$ where $n = 1$, leads to a *slope* $= -0.0591p = -0.060$. Solving for p leads to $p = 1.02 = 1$. A value of $r^2 = 1$ indicates that the data are linear. Because $E_{1/2}^A$, C_C^*, I_A, and I_{AC_p} are given, the terms in equation (1) can be tabulated as follows.

C_C^* (M)	$E_{1/2}^{AC_p}$ (V vs SCE)	$(I)_{\max} = (I)_{\max,AC_p}$	$E_{1/2}^{AC_p} - E_{1/2}^A$ (V)	$-0.0591\times$ $\log C_C^*$	$-0.0591\times$ $\log(I_A/I_{AC})$	K_c
10^{-3}	-2.15	2.03	-0.25	-01773	1.474×10^{-3}	1.80×10^7
10^{-2}	-2.21	2.02	-0.31	-0.1182	1.601×10^{-3}	1.87×10^7
10^{-1}	-2.27	2.04	-0.37	-0.0591	1.348×10^{-3}	1.92×10^7

K_c is calculated and tabulated for each C_C^* value. $K_c^{avg} = 1.9 \times 10^7 = \frac{C_{AC_p}}{C_A C_C^p}$ in the last column in the table. For $K_c^{avg} >> 1$, formation of the complex of A and C is favored with $p = 1$ to form AC_1.

Thus, a diffusion coefficient and equilibrium constant, both thermodynamic quantities, can be calculated from the data. Species A is probably an alkali metal given the charge of A and the small cavity in C. Crown ethers like C are notably good ligands for alkali metal ions in aqueous solution.

Problem 8.4© The diffusion current constant is calculated from equation (8.1.6).

$$(I)_{\max} = \frac{(i_d)_{\max}}{m^{2/3}t_{\max}^{1/6}C_O^*} = 708nD_O^{1/2} \tag{8.1.6}$$

The results are tabulated as follows.

$i_{d,\max}$	(μA)	3.9	6.5
$t_{\max}$	(s)	3.8	3.0
m	(mg/s)	1.85	1.85
$(I)_{\max}$	$(cm/s^{1/2})$	8.3	14.4

Expected reactions for the first and second wave are shown in Figure 6.4.1 in pH neutral electrolyte.

$$O_2 + 2H_2O + 2e \rightleftharpoons H_2O_2 + 2OH^-$$
$$H_2O_2 + 2e \rightleftharpoons 2OH^-$$

From the reactions, one would expect that the ratio of the diffusion current constants would be 2. The first wave is set by two electrons transferred; the second wave is set by four electrons transferred because both reduction processes are occurring at the more extreme potential. For the data provided here, however, the ratio is 14.4/8.3 = 1.7. The limiting current for both waves is diffusion limited, so there are no heterogeneous effects to consider in examining the ratio of diffusion current constants. Homogeneous reactions can affect the mass transport limited current. The first wave is too high relative to the second. If the second is $4e$ overall, then the first is more than $2e$. The explanation normally accepted is that H_2O_2 disproportionates slightly, regenerating O_2, with the effect that overall $n > 2$ for the first wave.

Based on the discussion above, the height of the second wave is used to calculate the diffusion coefficient of oxygen. For the second wave, n is firmly 4.

$$D_O = \left[\frac{(I)_{\max}}{708n}\right]^2 = \left[\frac{14.4}{708 \times 4}\right]^2 = 2.5 \times 10^{-5} \text{ cm}^2/\text{s}$$

=====

Problem 8.7© **(a).** A reversible sampled-current voltammetric wave is described by equation (6.2.8).

$$i(t) = \frac{nFAD_O^{1/2}C_O^*}{\sqrt{\pi t}(1 + \xi\theta)} \tag{6.2.8}$$

where

$$\theta = \exp\left[\frac{nF}{RT}\left(E - E^{0'}\right)\right] = \frac{C_O(0,t)}{C_R(0,t)} \tag{6.2.4}$$

$\xi = \sqrt{D_O/D_R}$. One can write

$$\frac{di}{dE} = \frac{di}{d\theta}\frac{d\theta}{dE} \tag{1}$$

$$\frac{di}{d\theta} = \frac{nFAD_O^{1/2}C_O^*}{\sqrt{\pi t}}\left(-\frac{\xi}{(1+\xi\theta)^2}\right) \tag{2}$$

$$\frac{d\theta}{dE} = \frac{nF}{RT}\exp\left[\frac{nF}{RT}\left(E - E^{0'}\right)\right] = \frac{nF\theta}{RT} \tag{3}$$

Combining equations (1) - (3) and recognizing that t is the sampling time τ leads to

$$\frac{di\left(\tau\right)}{dE} = -\frac{nFAD_O^{1/2}C_O^*\xi}{\sqrt{\pi\tau}(1+\xi\theta)^2}\frac{nF\theta}{RT} = -\frac{n^2F^2AD_O^{1/2}C_O^*}{RT\sqrt{\pi\tau}}\frac{\xi\theta}{(1+\xi\theta)^2} \tag{4}$$

(**b**). From equation (8.4.7)

$$\delta i = \frac{nFAD_O^{1/2}C_O^*}{\sqrt{\pi\left(\tau-\tau'\right)}}\left[\frac{P_A\left(1-\sigma^2\right)}{\left(\sigma+P_A\right)\left(1+P_A\sigma\right)}\right] \tag{8.4.7}$$

where

$$P_A = \xi\exp\left[\frac{nF}{RT}\left(E+\frac{\Delta E}{2}-E^{0'}\right)\right] = \xi\theta\exp\left[\frac{nF}{RT}\frac{\Delta E}{2}\right] = \xi\theta\sigma \tag{8.4.4}$$

$$\sigma = \exp\left[\frac{nF}{RT}\frac{\Delta E}{2}\right] \tag{8.4.5}$$

As $\Delta E \to 0$, the argument under the exponential in equation (8.4.5) becomes sufficiently small that $\lim\limits_{x\to 0} e^x \to 1 + x$. Thus, $P_A \to \xi\theta$ as $\sigma \to 1 + \frac{nF}{RT}\frac{\Delta E}{2} = 1 + nf\frac{\Delta E}{2} \sim 1$. Upon substitution into equation (8.4.7),

$$\begin{aligned}
\delta i &= \frac{nFAD_O^{1/2}C_O^*}{\sqrt{\pi\left(\tau-\tau'\right)}}\left[\frac{P_A\left(1-\sigma^2\right)}{\left(\sigma+P_A\right)\left(1+P_A\sigma\right)}\right] \\
&= \frac{nFAD_O^{1/2}C_O^*}{\sqrt{\pi\left(\tau-\tau'\right)}}\left[\frac{\xi\theta\left[1-\left(1+nf\frac{\Delta E}{2}\right)^2\right]}{\left(1+nf\frac{\Delta E}{2}+\xi\theta\right)\left(1+\xi\theta\left(1+nf\frac{\Delta E}{2}\right)\right)}\right] \\
&= \frac{nFAD_O^{1/2}C_O^*}{\sqrt{\pi\left(\tau-\tau'\right)}}\left[\frac{\xi\theta\left[-nf\Delta E-\left(nf\frac{\Delta E}{2}\right)^2\right]}{\left(1+nf\frac{\Delta E}{2}+\xi\theta\right)\left(1+\xi\theta\left(1+nf\frac{\Delta E}{2}\right)\right)}\right] \\
&\cong \frac{nFAD_O^{1/2}C_O^*}{\sqrt{\pi\left(\tau-\tau'\right)}}\left[\frac{\xi\theta\left[-nf\Delta E\right]}{\left(1+\xi\theta\right)\left(1+\xi\theta\right)}\right] \\
&\cong \frac{nFAD_O^{1/2}C_O^*}{\sqrt{\pi\left(\tau-\tau'\right)}}\left[\frac{\xi\theta\left[-nf\Delta E\right]}{\left(1+\xi\theta\right)^2}\right]
\end{aligned}$$

Rearrangement yields an expression which approaches equation (4) as $\Delta E \to 0$.

$$\frac{\delta i}{\Delta E} \cong -\frac{n^2F^2AD_O^{1/2}C_O^*}{RT\sqrt{\pi\left(\tau-\tau'\right)}}\left[\frac{\xi\theta}{(1+\xi\theta)^2}\right]$$

9 CONTROLLED-CURRENT TECHNIQUES

Problem 9.2© From the development in Section 9.4.2, the applied current will be expressed as

$$i(t) = i_f + S_{t_f}(t)(i_r - i_f)$$

The Laplace transform yields

$$\bar{i}(s) = \frac{i_f}{s} + \frac{\exp[-st_f]}{s}(i_r - i_f) \tag{1}$$

If equation (9.2.6) is limited to the case of $C_R^* = 0$ and $x = 0$, and if equation (1) is then substituted, equation (2) results.

$$\overline{C_R}(0, s) = \frac{i_f + \exp[-st_f](i_r - i_f)}{nFAD_R^{1/2}s^{3/2}} \tag{2}$$

This inverts as

$$C_R(0, t) = \frac{2}{nFAD_R^{1/2}\pi^{1/2}}\left[i_f t^{1/2} + S_{t_f}(t)(i_r - i_f)(t - t_f)^{1/2}\right]$$

The reverse transition occurs when $C_R(0, t) = 0$ and $t = t_f + \tau_r$. At that time,

$$i_f(t_f + \tau_r)^{1/2} + (i_r - i_f)\tau_r^{1/2} = 0$$

For the case where $\tau_r = t_f$,

$$i_f(2t_f)^{1/2} + (i_r - i_f)t_f^{1/2} = \left(\sqrt{2} - 1\right)i_f + i_r = 0$$

This is only true if $i_r = -\left(\sqrt{2} - 1\right)i_f = -0.414i_f$, or $i_f/i_r = -2.42$.

$$=====$$

Problem 9.4© **(a)** Laplace transforms are used to derive the Sand equation for the case where the current $i(t)$ increases with time as the square root,

$$i(t) = \beta t^{1/2} \tag{1}$$

The Laplace transform is

$$\bar{i}(s) = \frac{\beta \pi^{1/2}}{2s^{3/2}} \tag{2}$$

Substitute equation (2) into equation (9.2.5) and set $x = 0$.

$$\bar{C}_O(x, s) = \frac{C_O^*}{s} - \frac{\bar{i}(s)}{nFAD_O^{1/2}s^{1/2}}\exp\left[-\frac{s}{D_O^{1/2}}x\right] \tag{9.2.5}$$

Electrochemical Methods: Fundamentals and Applications, Third Edition, Student Solutions Manual. Cynthia G. Zoski and Johna Leddy.
© 2025 John Wiley & Sons Ltd. Published 2025 by John Wiley & Sons Ltd.

Chapter 9   CONTROLLED-CURRENT TECHNIQUES

$$
\begin{aligned}
\bar{C}_O\left(0,s\right) &= \frac{C_O^*}{s} - \frac{\bar{i}\left(s\right)}{nFAD_O^{1/2}s^{1/2}}\exp\left[-\frac{s}{D_O^{1/2}}x\right]_{x=0} \\
&= \frac{C_O^*}{s} - \frac{\bar{i}\left(s\right)}{nFAD_O^{1/2}s^{1/2}} \\
&= \frac{C_O^*}{s} - \frac{\beta\pi^{1/2}}{2s^{3/2}}\frac{1}{nFAD_O^{1/2}s^{1/2}} \\
&= \frac{C_O^*}{s} - \frac{\beta\pi^{1/2}}{2nFAD_O^{1/2}s^2}
\end{aligned}
\tag{3}
$$

Transition time τ occurs when $C_O\left(0,t\right)=0$, which is also when $\bar{C}_O\left(0,s\right)=0$.

$$
0 = \frac{C_O^*}{s} - \frac{\beta\pi^{1/2}}{2nFAD_O^{1/2}s^2}
$$

Inversion yields

$$
0 = C_O^* - \frac{\beta\pi^{1/2}t}{2nFAD_O^{1/2}}
$$

Set $t=\tau$. On rearranging,

$$
\frac{\beta\tau}{C_O^*} = \frac{2nFAD_O^{1/2}}{\pi^{1/2}}
$$

For electrolysis with a current that increases with the square root of time ($i\left(t\right) = \beta t^{1/2}$), the transition time τ (rather than $\tau^{1/2}$) is proportional to C_O^*.

(b) Consider the reaction sequence

$$
\begin{aligned}
\text{O} + \text{n}_1\text{e} &\rightleftarrows \text{X} & E_1^{0\prime} & \tag{4} \\
\text{X} + \text{n}_2\text{e} &\rightleftarrows \text{R} & E_2^{0\prime} & \tag{5}
\end{aligned}
$$

For programmed current of $i\left(t\right) = \beta t^{1/2}$ (equation (1)), where $E_1^{0\prime} > E_2^{0\prime}$ and $E_1^{0\prime}$ and $E_2^{0\prime}$ differ sufficiently, two transition times are observed. For $t < \tau_1$, the current is entirely consumed by reaction (4), electrolysis of O to X. After τ_1, conversion of O to X continues but is insufficient to sustain the current demand of $i\left(t\right) = \beta t^{1/2}$. X is then electrolyzed to R to the extent required to sustain the current. The total current $i = i_1 + i_2$ is generated by simultaneous reactions (4) and (5) that generate currents i_1 and i_2, respectively.

Before τ_1 ($t < \tau_1$, $i_2 = 0$), only $\text{O} + \text{n}_1\text{e} \rightleftarrows \text{X}$ generates current, where $i = i_1$. The transition time is as considered in part (a), where $n = n_1$ and $\tau = \tau_1$.

$$
\tau_1 = \frac{2n_1FAD_O^{1/2}C_O^*}{\beta\sqrt{\pi}}
\tag{6}
$$

And from equation (3),

$$\bar{\imath}_1(s) = n_1 FAD_O^{1/2} s^{1/2} \left[\frac{C_O^*}{s} - \bar{C}_O(0, s) \right] \tag{7}$$

After τ_1 ($t > \tau_1$), flux of both O + n_1e $\rightleftarrows$ X and X + n_2e $\rightleftarrows$ R generate current. Once $t > \tau_1$ all the O that reaches the electrode is electrolyzed to X. Then, X is converted to R to the extent that satisfies the current demand of $i(t) = \beta t^{1/2}$. Generation of R satisfies the required $i_2(t)/n_2 FA$. From equation (9.2.6) for $C_R^* = 0$, $x = 0$, and $n = n_2$,

$$\bar{C}_R(0, s) = + \frac{\bar{\imath}_2(s)}{n_2 FAD_R^{1/2} s^{1/2}}$$

$$\bar{\imath}_2(s) = n_2 FAD_R^{1/2} s^{1/2} \bar{C}_R(0, s) \tag{8}$$

From equations (7) and (8), the total current $i(t)$ is defined at all time.

$$\bar{\imath}(s) = \bar{\imath}_1(s) + \bar{\imath}_2(s)$$

$$\bar{\imath}(s) = n_1 FAD_O^{1/2} s^{1/2} \left[\frac{C_O^*}{s} - \bar{C}_O(0, s) \right] + n_2 FAD_R^{1/2} s^{1/2} \bar{C}_R(0, s) \tag{9}$$

At $t = \tau_1$, $\bar{C}_O(0, s) = 0$ and $\bar{C}_R(0, s) = 0$.

$$\bar{\imath}(s) = n_1 FAD_O^{1/2} \frac{C_O^*}{s^{1/2}} = \frac{\beta \pi^{1/2}}{2 s^{3/2}}$$

$$\frac{\beta \pi^{1/2}}{2 s^2} = \frac{n_1 FAD_O^{1/2} C_O^*}{s}$$

$$\tau_1 = \frac{2 n_1 FAD_O^{1/2} C_O^*}{\pi^{1/2} \beta} \tag{10}$$

This rederives equation (6) as equation (10) at $t = \tau_1$.

Once $t > \tau_1$, current is supported by two fluxes and R is present. Given semi-infinite conditions where flux of O to the surface equals flux of R from the surface, equation (11) applies. The derivation of general equation (11) is outlined below. $\xi = [D_O/D_R]^{1/2}$. Note that equation (11) applies to all O/R electrode reactions for semi-infinite linear diffusion conditions.

$$D_O^{1/2} C_O(0, t) + D_R^{1/2} C_R(0, t) = D_O^{1/2} C_O^* + D_R^{1/2} C_R^* \tag{11}$$

$$\xi \bar{C}_O(0, s) + \bar{C}_R(0, s) = \xi \frac{C_O^*}{s} + \frac{C_R^*}{s} \tag{12}$$

At $t = \tau_1 + \tau_2$, $\bar{C}_O(0, s) = 0$, $C_R^* = 0$ (no R initially present), and $\bar{C}_R(0, s) = \xi \frac{C_O^*}{s}$. For these conditions, substitution of equation (12) into equation (9) yields

$$\bar{i}(s) = n_1 F A D_O^{1/2} s^{1/2} \left[\frac{C_O^*}{s} \right] + n_2 F A D_R^{1/2} s^{1/2} \left[\xi \frac{C_O^*}{s} \right]$$

$$\frac{\beta \pi^{1/2}}{2 s^{3/2}} = n_1 F A D_O^{1/2} \frac{C_O^*}{s^{1/2}} + n_2 F A D_O^{1/2} \frac{C_O^*}{s^{1/2}}$$

$$\frac{\beta \pi^{1/2}}{2 s^2} = \frac{n_1 F A D_O^{1/2} C_O^*}{s} + \frac{n_2 F A D_O^{1/2} C_O^*}{s}$$

$$\frac{\beta \pi^{1/2}}{2} \left(\tau_1 + \tau_2 \right) = \left(n_1 + n_2 \right) F A D_O^{1/2} C_O^*$$

$$\tau_1 + \tau_2 = \frac{2 \left(n_1 + n_2 \right) F A D_O^{1/2} C_O^*}{\beta \pi^{1/2}} \tag{13}$$

τ_2 is found by subtracting equation (10) for τ_1 from equation (13).

$$\tau_2 = \frac{2 n_2 F A D_O^{1/2} C_O^*}{\beta \pi^{1/2}} \tag{14}$$

For $n_1 = n_2$ and $\tau_1 = \tau_2$.

General equation (11) is given as equation (5.4.27). in *Electrochemical Methods*, 2nd Edition (2001). The derivation by Laplace transform is as follows. Equation (11) is generally applicable provided (a) the system is for semi-infinite linear diffusion; and (b) flux of O to the electrode surface ($x = 0$) equals flux of R from the surface. Equations (6.1.6) and (6.1.8) are the general expressions for concentration profiles for O and R under semi-infinite conditions.

$$\bar{C}_O \left(x, s \right) = \frac{C_O^*}{s} + A(s) \exp \left[- \left(\frac{s}{D_O} \right)^{1/2} x \right] \tag{6.1.6}$$

$$\bar{C}_R \left(x, s \right) = \frac{C_R^*}{s} + B(s) \exp \left[- \left(\frac{s}{D_R} \right)^{1/2} x \right] \tag{6.1.8}$$

Flux of O to the electrode surface equal to flux of R from the electrode surface is specified and solved as shown.

$$D_O \frac{\partial \bar{C}_O \left(x, s \right)}{\partial x^2} \bigg|_{x=0} = -D_R \frac{\partial \bar{C}_R \left(x, s \right)}{\partial x^2} \bigg|_{x=0}$$

$$D_O A(s) \left[- \left(\frac{s}{D_O} \right)^{1/2} \right] = -D_R B(s) \left[- \left(\frac{s}{D_R} \right)^{1/2} \right]$$

$$D_O^{1/2} A(s) = -D_R^{1/2} B(s)$$

$$-\xi A(s) = B(s)$$

Then at $x = 0$, equations (6.1.6) and (6.1.8) define $\bar{C}_O(0,s)$ and $\bar{C}_R(0,s)$.

$$\begin{aligned} \bar{C}_O(0,s) &= \frac{C_O^*}{s} + A(s) \\ \bar{C}_R(0,s) &= \frac{C_R^*}{s} + B(s) = \frac{C_R^*}{s} - \xi A(s) \end{aligned}$$

Solve for $A(s)$ and rearrange.

$$\begin{aligned} A(s) &= \bar{C}_O(0,s) - \frac{C_O^*}{s} = -\frac{1}{\xi}\left(\bar{C}_R(0,s) - \frac{C_R^*}{s}\right) \\ D_O^{1/2}\left[\bar{C}_O(0,s) - \frac{C_O^*}{s}\right] &= D_R^{1/2}\left[\frac{C_R^*}{s} - \bar{C}_R(0,s)\right] \\ D_O^{1/2}\bar{C}_O(0,s) + D_R^{1/2}\bar{C}_R(0,s) &= D_O^{1/2}\frac{C_O^*}{s} + D_R^{1/2}\frac{C_R^*}{s} \qquad (15) \end{aligned}$$

Inversion of equation (15) yields the general result, equation (11).

=====

Problem 9.6© From equation (9.7.1), the charge on $C_{inj} = 1\ nF$ set by a 10 V battery is found.

$$\Delta q = C_{inj} \times V = 10^{-9}\ \text{F} \times 10\ \text{V} = 10^{-8}\ \text{C}$$

When Δq is distributed over $C_{inj} = 1$ nF and $C_d = 1\ \mu$F, the charge is conserved such that

$$\Delta q = q_{inj} + q_d = 10^{-8}\ \text{C}$$

Also, the voltage drop across the two capacitors must be equal. Thus,

$$\frac{q_{inj}}{C_{inj}} = \frac{q_d}{C_d} = \frac{q_{inj}}{10^{-9}\ \text{F}} = \frac{q_d}{10^{-6}\ \text{F}}$$

Solution of two equations in two unknowns yields $q_d = 9.99 \times 10^{-9}$ C and $q_{inj} = 9.99 \times 10^{-12}$ C. Thus, essentially all of the charge is delivered from C_{inj} to C_d.

The total capacitance in the Figure 9.9.1 circuit is found as follows.

$$\begin{aligned} \frac{1}{C_T} &= \frac{1}{C_{inj}} + \frac{1}{C_d} = 10^9 + 10^6 \approx 10^9\ \text{F}^{-1} \\ C_T &\approx 10^{-9}\ \text{F} \end{aligned}$$

Chapter 9   CONTROLLED-CURRENT TECHNIQUES

The time constant $\tau = R_s C_T \approx 100\ \Omega \times 10^{-9}\ \text{F} \approx 10^{-7}\ s$. From equation (1.2.6), the current for charging C_d drops to 5% of its initial value at $t = 3\tau$ and 1% of its initial value at $t = 5\tau$. Thus, C_d is 95% charged in $\approx 3 \times 10^{-7}$ s or $> 99\%$ charged in $\approx 5 \times 10^{-7}$ s.

=====

Problem 9.7©   To derive equation (9.7.7) from equation (9.7.6) leads to

$$\eta\left(t\right) = \eta\left(t = 0\right) - \frac{1}{R_{ct}C_d} \int_0^t \eta\left(t\right) dt \qquad (9.7.6)$$

The Laplace transform of equation (9.7.6) is

$$\overline{\eta}\left(s\right) = \frac{\eta\left(t = 0\right)}{s} - \frac{1}{R_{ct}C_d}\frac{\overline{\eta}\left(s\right)}{s}$$

which on rearrangement becomes

$$\overline{\eta}\left(s\right) = \frac{\eta\left(t = 0\right)}{s\left[1 + \frac{1}{R_{ct}C_d s}\right]} = \frac{\eta\left(t = 0\right)}{s + \frac{1}{R_{ct}C_d}}$$

From Table A.1.1, inversion leads to

$$\eta\left(t\right) = \eta\left(t = 0\right) \exp\left[-\frac{t}{R_{ct}C_d}\right] \qquad (9.7.7)$$

=====

Problem 9.9©   Barker studied the reactions of solvated electrons in acid solution. Electrons were ejected from the electrode using a light pulse and became solvated (as e_{aq}) in a part of the solution very near the electrode. The loss of charge from the electrode caused electrode potential to shift positively from the initial potential.

When no N_2O was present, e_{aq} could not react, and diffused back to the electrode where the electrons were collected. Thus, the potential of the electrode drifted monotonically negatively, toward the potential preceding the flash.

When N_2O is present in solution, e_{aq} reacted with N_2O to generate $OH^\bullet$, which could be reduced at the electrode. After the flash, the early part of the transient with N_2O present matched that of the transient with N_2O absent. The ejection of electrons took place in the time before the peak, and e_{aq} began to be re-collected at the electrode after the peak. In the presence of N_2O, the transient did not continue toward negative potentials, but turned positive. This branch reflects the withdrawal of additional electrons from the electrode as $OH^\bullet$ diffused there to be reduced. Evidently, the reaction to produce $OH^\bullet$ required 30 ns or so.

It is noted that solvated electrons can be generated in water by radiolysis, but solvated electrons are not encountered in ordinary aqueous electrochemistry.

10 METHODS INVOLVING FORCED CONVECTION - HYDRODYNAMIC METHODS

Problem 10.1[©] Information for the Rotating Disc Electrode (RDE):

$r_1 = 0.20$ cm
$A = \pi r^2 = \pi \times (0.20 \text{ cm})^2 = 0.126 \text{ cm}^2$
$C_A^* = 10^{-2} \text{ M} = 10^{-5} \text{ mol/cm}^3$
$D_A = 5 \times 10^{-6} \text{ cm}^2/\text{s}$
$f = 100 \text{ rpm} = 100 \text{ rev/min} \times 1 \text{ min}/60 \text{ s} = 1.67 \text{ rev/s}$
$\omega = 2\pi f = 2\pi \times 1.67 \text{ s}^{-1} = 10.5 \text{ s}^{-1}$
$\nu = 0.01 \text{ cm}^2/\text{s}$

$$A + e \rightleftharpoons A^-$$

(a). From equation (10.2.8),

$$v_y = -0.51\omega^{3/2}\nu^{-1/2}y^2 \tag{10.2.8}$$

$$= -\frac{0.51 \times \left(10.5 \text{ s}^{-1}\right)^{3/2}}{\left(0.01 \text{ cm}^2/\text{s}\right)^{1/2}}y^2 = -\frac{174}{\text{cm} \times \text{s}}y^2 \tag{1}$$

From equation (10.2.9),

$$v_r = 0.51\omega^{3/2}\nu^{-1/2}ry \tag{10.2.9}$$

$$= \frac{0.51 \times \left(10.5 \text{ s}^{-1}\right)^{3/2}}{\left(0.01 \text{ cm}^2/\text{s}\right)^{1/2}}ry = \frac{174}{\text{cm} \times \text{s}}ry \tag{2}$$

At $y = 10^{-3}$ cm and $r = 0.2$ cm, equations 10.2.8 and 10.2.9 lead, respectively, to $\nu_y = -1.74 \times 10^{-4}$ cm/s and $v_r = 3.48 \times 10^{-2}$ cm/s.

(b). At the electrode surface, where $y = 0$, $\nu_y = \nu_r = 0$.

(c). The values U_o, $i_{l,c}$, m_A, δ_A, and the Levich constant are calculated as follows.

From equation (10.2.10),

$$U_o = -0.88447 \left(\omega\nu\right)^{1/2} \tag{10.2.10}$$

$$= -0.88447 \times \left(\frac{10.5}{\text{s}} \times \frac{0.01 \text{ cm}^2}{\text{s}}\right)^{1/2} = -0.29 \text{ cm/s}$$

Electrochemical Methods: Fundamentals and Applications, Third Edition, Student Solutions Manual. Cynthia G. Zoski and Johna Leddy.
© 2025 John Wiley & Sons Ltd. Published 2025 by John Wiley & Sons Ltd.

Chapter 10   HYDRODYNAMIC METHODS

From equation (10.2.21),

$$i_{l,c} = 0.62nFAD_O^{2/3}\omega^{1/2}\nu^{-1/6}C_O^* \tag{10.2.21}$$

$$= \frac{0.62 \times 1\times\frac{96485\text{ C}}{\text{mol}}\times0.126\text{ cm}^2\times \left(\frac{5\times10^{-6}\text{ cm}^2}{\text{s}}\right)^{2/3} \times \left(10.5\text{ s}^{-1}\right)^{1/2} \times \frac{10^{-5}\text{ mol}}{\text{cm}^3}}{(0.01\text{ cm}^2/\text{s})^{1/6}}$$

$$= 154\ \mu\text{A}$$

From equation (10.2.22),

$$m_A = \frac{i_{l,c}}{nFAC_O^*}$$

$$= \frac{154 \times 10^{-6}\text{ A}}{(96485\text{ C/mol})(0.126\text{ cm}^2)(10^{-5}\text{ mol/cm}^3)} = 1.27 \times 10^{-3}\text{ cm/s}$$

From equation (10.2.23),

$$\delta_0 = \frac{D_A}{m_A}$$

$$= \frac{5 \times 10^{-6}\text{ cm}^2/\text{s}}{1.27 \times 10^{-3}\text{ cm/s}} = 3.94 \times 10^{-3}\text{ cm}$$

From page 418,

$$\text{Levich Constant} = \frac{i_{l,c}}{\omega^{1/2}C_A^*}$$

$$= \frac{154 \times 10^{-6}\text{ A}}{(10.5\text{ s}^{-1})^{1/2}(10^{-5}\text{ mol/cm}^3)} = 4.75\text{ As}^{1/2}\text{cm}^3/\text{mol}$$

=====

Problem 10.3© This problem is based on the data in Figure (10.2.8). From the Figure legend,
$f = 2500$ rpm$= 2500$ rev/min$\times 1$ min/60 s $= 41.67$ rev/s
$\omega = 2\pi f = 2\pi \times 41.67\text{ s}^{-1} = 262/\text{s}$
$\omega^{1/2} = 16.2\text{ s}^{-1/2}$
Au electrode, $A = 0.196\text{ cm}^2$
$C_{O_2}^* = 1.$ mM (saturated) $= 1. \times 10^{-6}\text{ mol/cm}^3$

$$O_2 + H_2O + 2e \rightleftharpoons HO_2^- + OH^- \qquad n = 2$$

(a). The D_{O_2} in 0.1 M NaOH is found from the $i - E$ curve in Figure (10.2.8a), where $i_{l,c} \approx 6.5 \times 10^{-4}$ A. From equation (10.2.21)

$$
\begin{aligned}
D_{O_2}^{2/3} &= \frac{i_{l,c}}{0.62nFA\omega^{1/2}\nu^{-1/6}C_{O_2}^*} \\
&= \frac{\left(6.5 \times 10^{-4}\ \text{A}\right)\left(0.01\ \text{cm}^2/\text{s}\right)^{1/6}}{(0.62)\,(2)\,(96485\ \text{C/mol})\,(0.196\ \text{cm}^2)\,(16.2\ \text{s}^{-1/2})\,(10^{-6}\ \text{mol/cm}^3)}
\end{aligned}
$$

$$
D_{O_2} = \left[7.94 \times 10^{-4}\ \frac{\text{cm}^{4/3}}{\text{s}^{2/3}}\right]^{3/2} = 2 \times 10^{-5}\ \text{cm}^2/\text{s}
$$

(b). The Koutecky Levich equation (10.2.39) shows that a graph of $1/i$ versus $\omega^{-1/2}$ leads to an intercept of $1/i_K$. From Figure 10.2.8, the intercept at 0.75 V is $i^{-1} = i_K^{-1} \approx 1.2$ mA^{-1}. This yields $i_K = 8.3 \times 10^{-4}$ A. From equation (10.2.38),

$$
\begin{aligned}
k_f(E) &= \frac{i_K}{nFAC_{O_2}^*} \\
&= \frac{8.3 \times 10^{-4}\ \text{A}}{(2)\,(96485\ \text{C/mol})\,(0.196\ \text{cm}^2)\,(10^{-6}\ \text{mol/cm}^3)} \\
&= 2 \times 10^{-2}\ \text{cm/s at 0.75V}
\end{aligned}
$$

$$=====$$

Problem 10.5[©] This problem is based on Figure 10.8.2, which features RRDE voltammograms for the reduction.

$$
\text{Fe}\,(\text{CN})_6^{3-} + e \rightleftharpoons \text{Fe}\,(\text{CN})_6^{4-}
$$

The data provided are as follows.

$f = 48.6$ rev/s
$\omega = 2\pi f = 2\pi(48.6\ \text{s}^{-1}) = 305\ \text{s}^{-1}$
$\omega^{1/2} = 17.5\ \text{s}^{-1/2}$
$C^* = 5.0\ \text{mM} = 5 \times 10^{-6}\ \text{mol/cm}^3$
$r_2 = 0.188\ \text{cm} = $ inner radius
$r_3 = 0.325\ \text{cm} = $ outer radius
$r_3^3 - r_2^3 = (0.325\ \text{cm})^3 - (0.188\ \text{cm})^3 = 2.77 \times 10^{-2}\ \text{cm}^3$
$\pi\,(r_3^3 - r_2^3)^{2/3} = 0.287\ \text{cm}^2$
$\nu = 0.01\ \text{cm}^2/\text{s}$
$\nu^{-1/6} = 2.15\ \text{cm}^{-2/6}/\text{s}^{-1/6}$

Chapter 10   HYDRODYNAMIC METHODS

(a). From Figure 10.8.2,

	i_D (μA)	$i_{R,l,c}$ (μA)
Curve (1)	0	≈ 1380
Curve (2)	302	≈ 1200

From equation (10.3.22),

$$
\begin{aligned}
N &= -\left(\frac{i_{R,l} - i_{R,l}^o}{i_D}\right) \quad\quad\quad (10.3.22)\\
&= -\left(\frac{1200\ \mu\text{A} - 1380\ \mu\text{A}}{302\ \mu\text{A}}\right) = 0.60
\end{aligned}
$$

From equation (10.3.8), corresponding to $i_d = 0$,

$$
\begin{aligned}
D_O^{2/3} &= \frac{i_{R,l,c}}{0.62 n F \pi \left(r_3^3 - r_2^3\right)^{2/3} \omega^{1/2} \nu^{-1/6} C_O^*}\\
&= \frac{1.38 \times 10^{-3}\ A}{0.62 \times 1 \times \frac{96485\ \text{C}}{\text{mol}} \times 0.287\ \text{cm}^2 \times 17.5\ \text{s}^{-1/2} \times \frac{2.15\ \text{cm}^{-2/6}}{\text{s}^{-1/6}} \times \frac{5 \times 10^{-6}\ \text{mol}}{\text{cm}^3}}\\
&= 4.27 \times 10^{-4}\ \text{cm}^{4/3}/\text{s}^{2/3}
\end{aligned}
$$

Or,

$$
D_O = \left(4.27 \times 10^{-4}\ \text{cm}^{4/3}/\text{s}^{2/3}\right)^{3/2} = 8.8 \times 10^{-6}\ \text{cm}^2/\text{s}
$$

(b). From equation (10.2.21),

$$
\begin{aligned}
\frac{i_{l,c}}{\omega^{1/2}} &= 0.62 n F A D_O^{2/3} \upsilon^{-1/6} C_O^*\\
&= \frac{302\ \mu\text{A}}{17.5\ \text{s}^{-1/2}} = 17.3\ \mu\text{A s}^{1/2}
\end{aligned}
$$

(c). Values for $i_{D,l,c}$ and $i_{R,l,c}$ at 5000 rpm are found through the proportionality of equation (10.2.21) between i_l and $\omega^{1/2}$. Data from Figure 10.8.2 are used.

$f = 5000$ rev/min$= 5000$ rev/min$\times 1$ min $/60$ s $= 83.3$ rev/s
$\omega = 2\pi f = 2\pi \times (83.3 \text{ rev/s}) = 523.6 \text{ s}^{-1}$
$\omega^{1/2} = 22.9 \text{ s}^{-1/2}$

$$i_{D,l,c} = 302 \ \mu A \times \left(\frac{22.9 \text{ s}^{-1/2}}{17.5 \text{ s}^{-1/2}} \right) = 395 \ \mu A$$

$$i_{R,l,c} \left(i_D = 0 \right) = 1380 \ \mu A \times \left(\frac{22.9/s^{1/2}}{17.5/s^{1/2}} \right) = 1.81 \text{ mA}$$

$$i_{R,l,c} \left(i_D = i_{D,l,c} \right) = 1200 \ \mu A \times \left(\frac{22.9/s^{1/2}}{17.5/s^{1/2}} \right) = 1.57 \text{ mA}$$

=====

Problem 10.7$^{\copyright}$ At $\omega = 0$, the Cottrell equation (6.1.12) holds, and for $\omega > 0$, the Levich equation (10.2.21) holds. The ratio yields

$$\frac{i_d \left(t \right) t^{1/2}}{i_l / \omega^{1/2}} = \frac{nFAD_O^{1/2}C_O^* \pi^{-1/2}}{0.62nFAD_O^{2/3}\nu^{-1/6}C_O^*} = 0.91\nu^{1/6}D_O^{-1/6}$$

Experimentally, this is the ratio of the slope of a Cottrell plot (i versus $t^{-1/2}$) to the slope of a Levich plot (i_l versus $\omega^{1/2}$). This is very useful in data analysis as it yields the diffusion coefficient without knowledge of n, A, and C_O^*. Kinematic viscosity ν is needed but 0.01 cm^2/s is applicable across a wide range of common electrolyte solutions.

Three assumptions apply: the electrolysis is mass transport limited; the same double layer phenomena are applicable to both measurements; and there are no heterogeneous or homogeneous kinetic effects on the time scale of the measurement. The method is reasonable for cases where there are no kinetic effects and the chronoamperometry data are taken in a quiescent solution for $t > 4R_uC_{dl}$, so that nonfaradaic charging current for double layer formation does not contribute to measured $i_d \left(t \right)$.

Chapter 10   HYDRODYNAMIC METHODS

Problem 10.10$^{©}$ The reader is asked to show that (10.2.40) and (10.2.41) lead to (10.2.39) with only the added recognition of $i_{l,c} = nFAm_OC_O^*$. The starting relationships (as revised in the errata for the text) are

$$\frac{i}{nFA} = m_O\left[C_O^* - C_O(y=0)\right] = k'C_O(y=0) \tag{10.2.40}$$

The highest possible rate of the interfacial reaction is reached when $C_O(y=0)$ approaches C_O^*. This rate defines a corresponding current, i_K, given in equation (10.2.41).

$$\frac{i_K}{nFA} = k'C_O^* \tag{10.2.41}$$

Using only the first relationship in (10.2.40), one has

$$\frac{i}{nFA} = m_O\left[C_O^* - C_O(y=0)\right] \tag{1}$$

The definition of the limiting current $i_{l,c}$ is rearranged.

$$\frac{i_{l,c}}{nFA} = m_OC_O^* \tag{2}$$

Divide equation (1) by equation (2).

$$\frac{i}{i_{l,c}} = 1 - \frac{C_O(y=0)}{C_O^*}$$

$$\frac{C_O(y=0)}{C_O^*} = 1 - \frac{i}{i_{l,c}} \tag{3}$$

From equation (10.2.40),

$$\frac{i}{nFA} = k'C_O(y=0)$$

Substitute k' from equation (10.2.41):

$$\frac{i}{nFA} = \frac{i_K}{nFAC_O^*}C_O(y=0)$$

$$i = i_K\left(\frac{C_O(y=0)}{C_O^*}\right)$$

Then, from equation (3),

$$i = i_K\left(1 - \frac{i}{i_{l,c}}\right)$$

$$\frac{i}{i_K} = 1 - \frac{i}{i_{l,c}}$$

Solve for i.

$$\frac{i}{i_K} + \frac{i}{i_{l,c}} = 1 = i\left(\frac{1}{i_K} + \frac{1}{i_{l,c}}\right)$$

$$i = \left(\frac{1}{i_K} + \frac{1}{i_{l,c}}\right)^{-1}$$

Inversion yields equation (10.2.39).

$$\frac{1}{i} = \frac{1}{i_K} + \frac{1}{i_{l,c}} \tag{10.2.39}$$

11 ELECTROCHEMICAL IMPEDANCE SPECTROSCOPY AND AC VOLTAMMETRY

Problem 11.1[©] The analogous problem of converting a series circuit to its parallel equivalent is outlined in the first edition on page 348. For components in parallel, the reciprocal of the total impedance is the sum of the reciprocals of the individual impedances. See Figure 11.2.10.

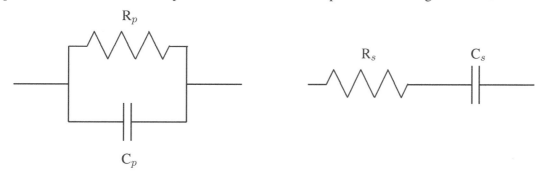

For a resistor, the impedance is R; for a capacitor, it is $[j\omega C]^{-1}$.

For the parallel network,

$$\frac{1}{\mathbf{Z}} = \frac{1}{R_p} + j\omega C_p \tag{1}$$

For components in series, the total impedance is the sum of the individual impedances.

$$\mathbf{Z} = R_s + \frac{1}{j\omega C_s} \tag{2}$$

Or, upon taking the reciprocal and noting $1/j = [-\sqrt{-1}\sqrt{-1}]/\sqrt{-1} = -\sqrt{-1} = -j$, equation (2) can be written as

$$\frac{1}{\mathbf{Z}} = \frac{1}{R_s + \frac{1}{j\omega C_s}} = \frac{\omega C_s}{\omega R_s C_s + \frac{1}{j}} = \frac{\omega C_s}{\omega R_s C_s - j}$$

Multiplying the numerator and denominator by the complement of the denominator yields

$$\frac{1}{\mathbf{Z}} = \frac{\omega C_s}{\omega R_s C_s - j} \times \frac{\omega R_s C_s + j}{\omega R_s C_s + j} = \frac{\omega C_s (\omega R_s C_s + j)}{(\omega R_s C_s)^2 + 1}$$

For simplification, let $W_s = (\omega R_s C_s)^2$.

$$\frac{1}{\mathbf{Z}} = \frac{R_s^{-1} W_s}{W_s + 1} + \frac{j\omega C_s}{W_s + 1} \tag{3}$$

Electrochemical Methods: Fundamentals and Applications, Third Edition, Student Solutions Manual. Cynthia G. Zoski and Johna Leddy.
© 2025 John Wiley & Sons Ltd. Published 2025 by John Wiley & Sons Ltd.

The final step is to recognize that the real parts in equation (3) correspond to the real parts in equation (1) and imaginary parts in the two equations also correspond. Thus,

$$\frac{1}{R_p} = \frac{W_s}{R_s(W_s+1)}$$

$$\omega C_p = \frac{\omega C_s}{W_s+1}$$

Or,

$$R_p = R_s\frac{W_s+1}{W_s}$$

$$C_p = C_s\frac{1}{W_s+1}$$

The conversion from a series to a parallel network develops similarly in that the reciprocal is taken of equation (3) and the real and imaginary parts are equated to the corresponding parts in equation (2). As outlined in the first edition, for $W_p = (\omega R_p C_p)^2$ this will yield

$$R_s = \frac{R_p}{W_p+1}$$

$$C_s = C_p\frac{W_p+1}{W_p}$$

=====

Problem 11.3$^{©}$ **(a).** The experimental method used here is described in Section 11.1. The impedance of the whole cell is measured as R_B and C_B in series. The simple equivalent circuit of $R_B C_B$ can be elaborated as in Figure 11.3.1 where R_u is in series with parallel components of C_d and the faradaic impedance, Z_f. The faradaic impedance is represented as a series RC network where the elements are R_s and C_s. If Z_f can be extracted from the measurables R_B and C_B, then R_s and C_s can be determined. The trick is to note that for resistors in series, the total resistance is the sum of the resistances; for capacitors in parallel, the total capacitance is the sum of the capacitances. R_u and C_d can be eliminated from the measurables by first considering a series circuit (to eliminate R_u) and then a parallel circuit (to eliminate C_d).

First, consider R_B which is composed of two components, R_u in series with the parallel element. As this is a series circuit, the measured resistance R_B can be expressed as $R_B = R_u + R'_B$ where R'_B is the resistance of the parallel element. Thus, the solution resistance can be eliminated as

$$R'_B = R_B - R_u \tag{4}$$

Second, this leaves a parallel circuit where C_d is in parallel with the faradaic impedance. The series values (R'_B and C_B) can be converted to parallel components following the equations developed in

Problem 11.1. For $W = (\omega R C)^2 = (\omega R'_B C_B)^2$,

$$R_p = R\left[\frac{W+1}{W}\right] = R'_B\left[\frac{W+1}{W}\right] \tag{5}$$

$$C_p = \frac{C}{W+1} = \frac{C_B}{W+1} \tag{6}$$

Then, the double layer capacitance is eliminated as

$$C'_p = C_p - C_d \tag{7}$$

Third, the faradaic impedance remains in a parallel arrangement. It remains to convert the parallel form to the series form. Equations were developed in Problem 11.1. For $W_p = \left(\omega R_p C'_p\right)^2$,

$$R_s = \frac{R_p}{1 + W_p} \tag{8}$$

$$C_s = C'_p \frac{1 + W_p}{W_p} \tag{9}$$

Finally, the phase angle is calculated from equation (11.3.35). Note that radians are converted to degrees by multiplying by $180°/\pi$.

$$\phi = \tan^{-1}\left[\frac{1}{\omega R_s C_s}\right] = \arctan\left[\frac{1}{\omega R_s C_s}\right] \tag{11.3.35}$$

Values and the corresponding equations are tabulated on the next page. $R_s = 10\ \Omega$; $C_d = 20.0\ \mu F$.

	eqn.		ω			
			49	100	400	900
R_B	(Ω)		146.1	121.6	63.3	30.2
C_B	(μF)		290.8	158.6	41.4	25.6
$R'_B = R_B - R_u$	(Ω)	(4)	136.1	111.6	53.3	20.2
$W = (\omega R'_B C_B)^2$			3.761	3.133	0.779	0.217
R_p	(Ω)	(5)	172.3	147.2	121.7	113.5
C_p	(μF)	(6)	61.1	38.4	23.3	21.0
$C'_p = C_p - Cd$	(μF)	(7)	41.1	18.4	3.3	1.0
$W_p = (\omega R_p C'_p)^2$			0.120	0.0732	0.0254	0.0113
R_s	(Ω)	(8)	153.8	137.2	118.7	112.2
C_s	(μF)	(9)	382.6	269.5	132.3	93.1
$[\omega R_s C_s]^{-1}$			0.347	0.271	0.159	0.106
ϕ	(rad)	(10.3.9)	0.334	0.264	0.158	0.106
ϕ	(deg)	(10.3.9)	19.1	15.1	9.05	6.07

Chapter 11   EIS AND AC VOLTAMMETRY

(b). Plots of R_s and C_s versus $\omega^{-1/2}$ will be linear and yield R_{ct} and σ.

$$R_s = R_{ct} + \frac{\sigma}{\omega^{1/2}} \qquad (11.3.25)$$

$$C_s = \frac{1}{\sigma\omega^{1/2}} \qquad (11.3.26)$$

In the plot, the markers are R_s ($\blacklozenge$) and C_s ($\circ$).

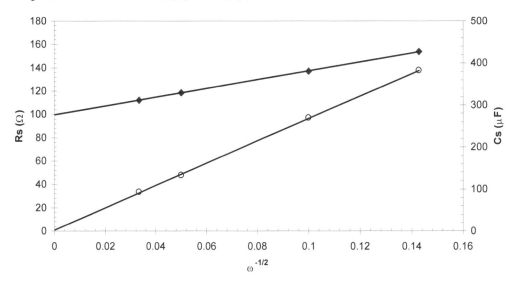

Relevant definitions are provided by equations (11.3.37), (3.4.6), and (11.3.33).

$$R_{ct} = \frac{RT}{Fi_0} \qquad (11.3.37)$$

$$i_0 = FAk^0 C_O^{*(1-\alpha)} C_R^{*\alpha} \qquad (3.4.6)$$

$$\sigma = \frac{RT}{F^2 A \sqrt{2}} \left[\frac{1}{\sqrt{D_O} C_O^*} + \frac{1}{\sqrt{D_R} C_R^*} \right] \qquad (11.3.33)$$

Regression analysis yields $R_s = 99.6 + 378/\omega^{1/2} = R_{ct} + \sigma/\omega^{1/2}$ and $C_s(F) = 2.66 \times 10^{-3}/\omega^{1/2} = 1/\sigma\omega^{1/2}$. Thus, $R_{ct} = 99.6 \ \Omega$ and $\sigma = 378$.

Equation (11.3.37) yields

$$i_0 = \frac{RT}{FR_{ct}} = \frac{0.02569 \text{ V}}{99.6 \ \Omega} = 2.58 \times 10^{-4} \text{ A}$$

It is given that $n = 1$, $A = 1 \text{ cm}^2$, and $C_O^* = C_R^* = C^* = 1.00 \times 10^{-6} \text{ mol/cm}^3$, such that from equation (3.4.6),

$$k^0 = \frac{i_0}{FAC^*}$$

$$= \frac{2.58 \times 10^{-4} \text{ A}}{96485 \text{ C/mol} \times 1 \text{ cm}^2 \times 1.00 \times 10^{-6} \text{ mol/cm}^3} = 2.67 \times 10^{-3} \text{ cm/s}$$

From equation (11.3.33),

$$\sqrt{D} = \frac{\sqrt{2}RT}{\sigma F^2 A C^*}$$

$$= \frac{\sqrt{2} \times 0.02569 \text{ V}}{378 \ \Omega/\text{s}^{1/2} \times 96485 \text{ C/mol} \times 1 \text{ cm}^2 \times 1.00 \times 10^{-6} \text{ mol/cm}^3}$$

$$= 1.0 \times 10^{-3} \text{ cm/s}^{1/2}$$

$$D = 1.0 \times 10^{-6} \text{ cm}^2/\text{s}$$

=====

Problem 11.5© The two equations to consider are

$$Z_{\text{Re}} = R_u + \frac{R_{ct}}{1 + (\omega C_d R_{ct})^2} \tag{11.4.9}$$

$$-Z_{\text{Im}} = \frac{\omega C_d R_{ct}^2}{1 + (\omega C_d R_{ct})^2} \tag{11.4.10}$$

Solve for ω by noting that

$$1 + (\omega C_d R_{ct})^2 = \frac{R_{ct}}{Z_{\text{Re}} - R_u} = \frac{\omega C_d R_{ct}^2}{-Z_{\text{Im}}}$$

Or, from the second two terms on the right hand side,

$$\omega = \frac{-Z_{\text{Im}}}{C_d R_{ct} [Z_{\text{Re}} - R_u]} \tag{10}$$

$$\omega C_d R_{ct} = \frac{-Z_{\text{Im}}}{Z_{\text{Re}} - R_u} \tag{11}$$

Equation (11.4.10) is rearranged to yield

$$(\omega C_d R_{ct})^2 + \frac{\omega C_d R_{ct}^2}{Z_{\text{Im}}} + 1 = 0$$

Substitution of equation (11) yields

$$\left(\frac{Z_{\text{Im}}}{Z_{\text{Re}} - R_u}\right)^2 - \frac{R_{ct}}{Z_{\text{Re}} - R_u} + 1 = 0$$

$$Z_{\text{Im}}^2 - R_{ct}[Z_{\text{Re}} - R_u] + [Z_{\text{Re}} - R_u]^2 = 0 \tag{12}$$

Note that algebraically

$$\left(Z_{\mathrm{Re}} - R_u - \frac{R_{ct}}{2}\right)^2 = [Z_{\mathrm{Re}} - R_u]^2 - R_{ct}[Z_{\mathrm{Re}} - R_u] + \frac{R_{ct}^2}{4} \tag{13}$$

Substitution of equation (13) into equation (12) yields equation (11.4.11).

$$\left(Z_{\mathrm{Re}} - R_u - \frac{R_{ct}}{2}\right)^2 + Z_{\mathrm{Im}}^2 = \frac{R_{ct}^2}{4} \tag{11.4.11}$$

=====

Problem 11.6[©] The redox species are bound directly to the electrode surface with surface coverages of O and R of $\Gamma_O(t)$ and $\Gamma_R(t)$ at initial concentrations of Γ_O^* and Γ_R^*. Because the redox centers are bound to the electrode and are interconvertible, the total concentration is always $\Gamma_O^* + \Gamma_R^*$. The initial concentrations, Γ_O^* and Γ_R^* can be determined by applying the appropriate dc potential. This is the problem of the electroactive monolayer outlined in Section 17.2.2. The equivalent circuit of Figure 17.3.3b is shown below. It is composed of R_u in series with a network, which is C_d in parallel with a linear RC circuit composed of R_a in series with C_a. Properties of the adsorbed species are characterized by R_a and C_a, where R_a is a charge transfer resistance. Support for this equivalent circuit can be found in the discussion in Section 17.3.3 and in E. Laviron, *J. Electroanal. Chem.*, **97** (1979) 135-149; E. Laviron, *J. Electroanal. Chem.*, **105** (1979) 35-42.

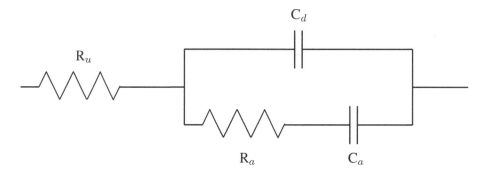

The faradaic current for adsorbed monolayers is defined by a modification of equation (17.2.6). For AC techniques, the anodic current is taken as positive.

$$\frac{\partial \Gamma_O(t)}{\partial t} = -\frac{\partial \Gamma_R(t)}{\partial t} = \frac{i}{nFA} \tag{14}$$

The development now follows the discussion in Section 11.3.2. Note that this circuit is identical to that in Figure 11.3.1, where the series components R_s and C_s are replaced by R_a and C_a. Thus, equations (11.3.1) to (11.3.4) apply, as listed below.

$$E = iR_a + \frac{q}{C_a} \tag{11.3.1}$$

$$\frac{dE}{dt} = R_a\frac{di}{dt} + \frac{i}{C_a} \tag{11.3.2}$$

$$i = I\sin\omega t \tag{11.3.3}$$

$$\frac{di}{dt} = I\omega\cos\omega t \tag{15}$$

$$\frac{dE}{dt} = R_aI\omega\cos\omega t + \frac{I}{C_a}\sin\omega t \tag{11.3.4}$$

Now, modification of equation (11.3.5) for surface concentrations yields $E = E\left[i, \Gamma_O(t), \Gamma_R(t)\right]$ and for the circuit above,

$$\frac{dE}{dt} = \left(\frac{\partial E}{\partial i}\right)\frac{di}{dt} + \left[\frac{\partial E}{\partial\Gamma_O(t)}\right]\frac{d\Gamma_O(t)}{dt} + \left[\frac{\partial E}{\partial\Gamma_R(t)}\right]\frac{d\Gamma_R(t)}{dt} \tag{16}$$

$$= R_{ct}\frac{di}{dt} + \beta_O\frac{d\Gamma_O(t)}{dt} + \beta_R\frac{d\Gamma_R(t)}{dt} \tag{17}$$

By analogy to equations (11.3.8) to (11.3.10),

$$R_{ct} = \left[\frac{\partial E}{\partial i}\right]_{\Gamma_O(t),\Gamma_R(t)} \tag{18}$$

$$\beta_O = \left[\frac{\partial E}{\partial\Gamma_O(t)}\right]_{i,\Gamma_R(t)} \tag{19}$$

$$\beta_R = \left[\frac{\partial E}{\partial\Gamma_R(t)}\right]_{i,\Gamma_O(t)} \tag{20}$$

It remains to evaluate the three derivative terms on the right hand side of equation (17). The first term, di/dt is specified by equation 11.3.3. The derivatives with respect to surface coverage are evaluated by Laplace transform of equation (14) and substitution of the initial conditions $\Gamma_O(0) = \Gamma_O^*$ and $\Gamma_R(0) = \Gamma_R^*$.

$$\bar{\Gamma}_O(s) = \frac{\Gamma_O^*}{s} + \frac{\bar{i}}{nFAs}$$

$$\bar{\Gamma}_R(s) = \frac{\Gamma_R^*}{s} - \frac{\bar{i}}{nFAs}$$

Because division by s corresponds to a simple integration in t-space, inversion yields the following where τ is the variable of integration. See also equation (A.1.17).

$$\Gamma_O(t) \;=\; \Gamma_O^* + \frac{1}{nFA} \int_0^t i(\tau)d\tau \tag{21}$$

$$\Gamma_R(t) \;=\; \Gamma_R^* - \frac{1}{nFA} \int_0^t i(\tau)d\tau \tag{22}$$

From equation (11.3.3), $i(\tau) = I \sin(\omega\tau)$. Thus,

$$
\begin{aligned}
\int_0^t i(\tau)d\tau \;&=\; I \int_0^t \sin(\omega\tau)\,d\tau \\
&=\; -I \frac{\cos\omega\tau}{\omega}\Big|_0^t \\
&=\; -\frac{I}{\omega}\left[\cos\omega t - 1\right] \\
&=\; \frac{I}{\omega}\left[1 - \cos\omega t\right]
\end{aligned}
$$

Thus, equations (21) and (22) become

$$\Gamma_O(t) \;=\; \Gamma_O^* + \frac{I}{nFA\omega}\left[1 - \cos\omega t\right] \tag{23}$$

$$\Gamma_R(t) \;=\; \Gamma_R^* - \frac{I}{nFA\omega}\left[1 - \cos\omega t\right] \tag{24}$$

And,

$$\frac{d\Gamma_O(t)}{dt} \;=\; \frac{I}{nFA} \sin\omega t \tag{25}$$

$$\frac{d\Gamma_R(t)}{dt} \;=\; -\frac{I}{nFA} \sin\omega t \tag{26}$$

Substitution of equations 11.3.3, (25), and (26) into equation (17) yields

$$
\begin{aligned}
\frac{dE}{dt} \;&=\; R_{ct}I\omega\cos\omega t + \beta_O \frac{I}{nFA}\sin\omega t - \beta_R \frac{I}{nFA}\sin\omega t \\
&=\; R_{ct}I\omega\cos\omega t + (\beta_O - \beta_R)\frac{I}{nFA}\sin\omega t
\end{aligned}
$$

Now, by comparison to equation (11.3.4), it becomes apparent that

$$R_a \;=\; R_{ct} \tag{27}$$

$$C_a \;=\; \frac{nFA}{\beta_O - \beta_R} \tag{28}$$

Equation (11.3.25) is the analogous expression for R_s when species are present in solution: $R_s = R_{ct} + \sigma/\sqrt{\omega}$. The second term describes mass transport. Here, the second term is absent, consistent with the adsorbed species not undergoing mass transport.

By analogy to equation (11.3.28),

$$\eta = E - E^{0'} = \frac{RT}{F}\left[\frac{\Gamma_O(t)}{\Gamma_O^*} - \frac{\Gamma_R(t)}{\Gamma_R^*} + \frac{i}{i_0}\right]$$

Derivatives based on equations (16), (18), (19), and (20) lead to the definitions of R_a, β_O, and β_R.

$$
\begin{aligned}
R_{ct} &= \left[\frac{\partial E}{\partial i}\right]_{\Gamma_O(t),\Gamma_R(t)} = \frac{RT}{Fi_0} \\
\beta_O &= \left[\frac{\partial E}{\partial \Gamma_O(t)}\right]_{i,\Gamma_R(t)} = \frac{RT}{F\Gamma_O^*} \\
\beta_R &= \left[\frac{\partial E}{\partial \Gamma_R(t)}\right]_{i,\Gamma_O(t)} = -\frac{RT}{F\Gamma_R^*}
\end{aligned}
$$

From equation (28),

$$C_a = \frac{nF^2A}{RT\left(\frac{1}{\Gamma_O^*} + \frac{1}{\Gamma_R^*}\right)}$$

For adsorbed layers, only R_a is needed to determine R_{ct} and the exchange current, which leads to k^0 through $i_0 = nFAk^0\Gamma_O^{*(1-\alpha)}\Gamma_R^{*\alpha}$, by analogy to equation (3.4.6).

From equation (11.2.11) and the discussion below it, the total impedance is expressed as

$$\mathbf{Z}(\omega) = Z_{\text{Re}} + jZ_{\text{Im}} \qquad (11.2.11)$$

For a series RC circuit, $Z_{\text{Re}} = R_a$ and $Z_{\text{Im}} = -1/\omega C_a$. Here, R_a and C_a are defined by equations (27) and (28). Equation (11.2.13) defines the phase angle ϕ through the relationship

$$
\begin{aligned}
\tan\phi &= -\frac{Z_{\text{Im}}}{Z_{\text{Re}}} \qquad (11.2.13) \\
&= \frac{1}{\omega R_a C_a} \\
&= \frac{\beta_O - \beta_R}{\omega R_{ct} nFA}
\end{aligned}
$$

$$\phi = \tan^{-1}\left[\frac{i_0\left(\frac{1}{\Gamma_O^*} + \frac{1}{\Gamma_R^*}\right)}{\omega nFA}\right] \qquad (29)$$

Chapter 11   EIS AND AC VOLTAMMETRY

As $R_{ct} \to 0$, $i_0 \to \infty$ and $\phi \to 90° = \pi/2$. This behavior is anticipated by the discussion in Section 17.3.3 in the text. Note that for species diffusing in solution as described on page 458 of the text, $\phi \le 45° = \pi/4$.

=====

Problem 11.9[©] From equation (11.3.35),

$$\phi = \tan^{-1}\left[\frac{\sigma/\omega^{1/2}}{R_{ct} + \sigma/\omega^{1/2}}\right] \tag{11.3.35}$$

where

$$\sigma = \frac{RT}{n^2 F^2 A \sqrt{2}}\left[\frac{1}{\sqrt{D_O}C_O^*} + \frac{1}{\sqrt{D_R}C_R^*}\right] \tag{11.3.33}$$

$$R_{ct} = \frac{RT}{F i_0} \tag{11.3.37}$$

$$i_0 = nFAk^0 C_O^{*(1-\alpha)} C_R^{*\alpha} \tag{3.4.6}$$

It is given that $k^0 = 2.2 \pm 0.3$ cm/s, $\alpha = 0.70$, $D_O = 1.02 \times 10^{-5}$ cm^2/s, $n = 1$, and $T = 295 \pm 2$ K. For $n = 1$, substitution of equations (11.3.33), (11.3.37), and (3.4.6) into equation (11.3.35) yields

$$
\begin{aligned}
\phi &= \tan^{-1}\left[\frac{\dfrac{RT}{n^2 F^2 A \sqrt{2\omega}}\left[\dfrac{1}{\sqrt{D_O}C_O^*} + \dfrac{1}{\sqrt{D_R}C_R^*}\right]}{\dfrac{RT}{FnFAk^0 C_O^{*(1-\alpha)} C_R^{*\alpha}} + \dfrac{RT}{n^2 F^2 A \sqrt{2\omega}}\left[\dfrac{1}{\sqrt{D_O}C_O^*} + \dfrac{1}{\sqrt{D_R}C_R^*}\right]}\right] \\[2em]
&= \tan^{-1}\left[\frac{k^0 C_O^{*(1-\alpha)} C_R^{*\alpha}\left[\dfrac{1}{\sqrt{D_O}C_O^*} + \dfrac{1}{\sqrt{D_R}C_R^*}\right]}{\sqrt{2\omega} + k^0 C_O^{*(1-\alpha)} C_R^{*\alpha}\left[\dfrac{1}{\sqrt{D_O}C_O^*} + \dfrac{1}{\sqrt{D_R}C_R^*}\right]}\right]
\end{aligned}
$$

Let $C_O^* = C_R^* = C^*$ and $D_R = D_O$.

$$\phi = \tan^{-1}\left[\frac{\dfrac{2k^0}{\sqrt{D_O}}}{\sqrt{2\omega} + \dfrac{2k^0}{\sqrt{D_O}}}\right]$$

It is given that $k^0 = 2.2 \pm 0.3$ cm/s, $\alpha = 0.70$, $D_O = 1.02 \times 10^{-5}$ cm^2/s, and $T = 295 \pm 2$ K, such that $k^0/\sqrt{D_O} = 688$ s$^{-1/2}$. For several decades of ω, ϕ is tabulated below.

$\omega/2\pi$	ω	$\phi(rad)$	$\phi(\deg)$	$\omega^{1/2}$	$\cot\phi$
10	62.8	0.7813	44.77	7.93	1.008
100	628	0.7727	44.27	25.07	1.026
1000	6283	0.7463	42.76	79.27	1.081
10000	62831	0.6718	38.49	250.66	1.258

For reversible reactions, $\phi = 45°$. For $k^0 = 2.2 \pm 0.3$ cm/s, the reaction will be reversible at low frequencies, as is consistent with the data in the table where $\phi \to 45°$ as ω decreases.

A plot of $\cot\phi = 1/\tan\phi$ versus $\omega^{1/2}$ is shown. Note that $E = E_{1/2} = E^{0'}$ when $D_O = D_R$; then, k^0 is the operative heterogeneous rate. For these conditions, equation (11.5.19) applies, and it simplifies as shown for $D = D_O = D_R$ where $\beta = 1 - \alpha$.

$$[\cot\phi]_{E_{1/2}} = 1 + \left[\frac{D_O^\beta D_R^\alpha}{2}\right]^{1/2} \frac{\omega^{1/2}}{k^0} \qquad (11.5.19)$$

$$= 1 + \frac{D^{1/2}}{\sqrt{2}k^0}\omega^{1/2}$$

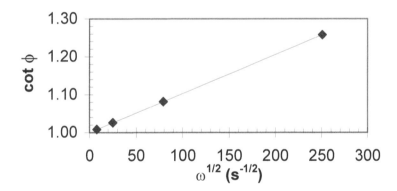

Regression yields $\cot\phi = 1.03 \times 10^{-3}\omega^{1/2} + 1.0000$. The *slope* $= \sqrt{D/2}/k^0$; for the values here, $\sqrt{D/2}/k^0 = 1.03 \times 10^{-3}$ s$^{1/2}$.

Consider Figure 11.3.4, which shows the real and imaginary vectors that define the response for a quasireversible electron transfer. The real vector, measured along the same vector as $\dot{E}_{ac}$ for a phase angle of $0°$, is $R_{ct} + \sigma/\omega^{1/2}$. The vector $90°$ out of phase defines the imaginary term, $\sigma/\omega^{1/2}$. The ratio of these two terms defines $\cot\phi$. From equation (11.3.35),

$$\cot\phi = \omega R_s C_s = \frac{R_{ct} + \sigma/\omega^{1/2}}{\sigma/\omega^{1/2}}$$

Thus, the ratio of a current measurement on the real axis made at $0°$ displacement with respect to $\dot{E}_{ac}$ and a second current measurement 90° out of phase (quadrature current) will yield $\cot\phi$. Note that this assumes effects from uncompensated solution resistance and double layer charging are negligible.

To make a good measurement of k^0, the frequency must be high enough that the measured value of ϕ must be less than 45°. As above, this condition is favored by higher frequency (faster measurements), greater than ≈ 10 kHz. Commercial instrumentation is available that has a frequency range of 20 MHz.

12 BULK ELECTROLYSIS

Problem 12.1[©]   The reaction under consideration is

$$\mathrm{M^{2+} + 2e \rightleftharpoons M}$$

For electrolysis at a rotating disk electrode, equation (1.3.10) is applicable.

$$i_l, c = nFAm_O C_O^* \tag{1.3.10}$$

It is given that $n = 2$, $C_O^* = 1.0 \times 10^{-5}$ mole/cm^3, $A = 10$ cm^2, and $i_l = 193$ mA.

$$m_{M^{2+}} = \frac{i_l}{nFAC_O^*} = \frac{193 \times 10^{-3}\ \mathrm{A}}{2 \times 96485\ C \times 10\ \mathrm{cm^2} \times 1.0 \times 10^{-5}\ \mathrm{mole/cm^3}} = 0.010\ \mathrm{cm/s}$$

The time required to electrolyze 99.9% of M^{2+} is found from equation (12.2.8), where $p = m_O A/V$.

$$t = -\frac{2.303}{p} \log\left[\frac{C_O^*(t)}{C_O^*(0)}\right] \tag{12.2.8}$$

Here, $V = 100$ cm^3.

$$
\begin{aligned}
t &= -\frac{2.3}{p} \log\left[\frac{C_O^*(t)}{C_O^*(0)}\right] \\
&= -\frac{2.3 \times 100\ \mathrm{cm^3}}{0.010\ \mathrm{cm/s} \times 10.\ \mathrm{cm^2}} \log[0.001] = 6900\ \mathrm{s} = 1.9\ \mathrm{hr}
\end{aligned}
$$

The charge required to electrolyze M^{2+} is calculated according to equation (12.2.12).

$$
\begin{aligned}
Q^0 &= nFN_0(0) = nFVC_O^*(0) \tag{12.2.12} \\
&= 2 \times 96485\ \mathrm{C/mole} \times 100\ \mathrm{cm^3} \times 1.0 \times 10^{-5}\ \mathrm{mole/cm^3} = 193\ \mathrm{C}
\end{aligned}
$$

=====

Problem 12.4[©]   The reactions under consideration

$$\mathrm{X^{2+} + 2e \overset{Hg}{\rightleftharpoons} X(Hg)} \quad E^{0\prime}_{X^{2+},X(Hg)} = -0.45\ \mathrm{V}\ vs\ \mathrm{SCE} \quad C_X^* = 1.0 \times 10^{-6}\ \mathrm{mole/cm^3}$$

$$\mathrm{Y^{2+} + 2e \overset{Hg}{\rightleftharpoons} Y(Hg)} \quad E^{0\prime}_{Y^{2+},Y(Hg)} = -0.70\ \mathrm{V}\ vs\ \mathrm{SCE} \quad C_Y^* = 3.0 \times 10^{-6}\ \mathrm{mole/cm^3}$$

It is also given that $m = 0.010$ cm/s, $V_S = 200$ cm^3, $A = 50$ cm^2, and $V_{Hg} = 100$ cm^2.

Electrochemical Methods: Fundamentals and Applications, Third Edition, Student Solutions Manual. Cynthia G. Zoski and Johna Leddy.
© 2025 John Wiley & Sons Ltd. Published 2025 by John Wiley & Sons Ltd.

Chapter 12   BULK ELECTROLYSIS

(a). This is a stirred system that should show steady-state mass transfer of X and Y to the mercury pool electrode. Most reductive amalgamation reactions are reversible, and one can assume that these reactions are reversible. No electrode reaction would occur between the positive background limit and about -0.4 V because there is nothing to oxidize or reduce in that range. There would be separate cathodic waves for X and Y, each located at $E_{1/2} \approx E^{0\prime}$. The wave for Y ($E_{1/2} = -0.70$ V) should be three times higher than that for X ($E_{1/2} = -0.45$ V). because $C_Y^* = 3C_X^*$. The limiting currents are readily calculated from $i_{l,c} = nFAm_OC_O^*$ (equation (1.3.10)), which gives 96 mA for X and 290 mA for Y. A quantitative sketch would look like the following.

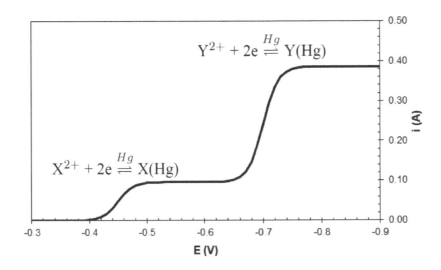

(b) Formation of an amalgam at a mercury electrode, $O + ne \rightleftharpoons R(Hg)$ occurs at a formal potential that differs from the value for the electrode reaction without amalgamation, $O + ne \rightleftharpoons R$. The formal potentials given in this problem are for the amalgamation reactions. The Nernstian relationship for the amalgam has the familiar form except C_R^* is in the mercury electrode. If potential E is applied, electrolysis adjusts the concentrations until a new Nernstian balance is established corresponding to E. Current flows to establish the new balance, but decays to zero as it is reached. Given the volume of solution V_s, the volume of the mercury V_{Hg}, an initial concentration of O, $C_{O,\text{initial}}^*$, and an initial concentration of R, $C_{R,\text{initial}}^* = 0$,

$$\text{moles O initially in solution} \quad = \quad V_s C_{O,\text{initial}}^* \tag{1}$$
$$\text{moles R initially in mercury} \quad = \quad 0 \tag{2}$$

In the long-time limit after potential E is applied, the concentration for O and R are given.

$$\lim_{t\to\infty} C_O^* \quad = \quad (\text{moles O at equilibrium in solution}) \, V_s^{-1} = C_{O,\text{initial}}^* \, (1-x)$$

$$\lim_{t\to\infty} C_R^* \quad = \quad (\text{moles R at equilibrium in mercury}) \, V_{Hg}^{-1} = \frac{C_{O,\text{initial}}^* x V_s}{V_{Hg}}$$

Substitution into equation (12.1.2) yields the relationship between E and x for the amalgam.

$$E = E^{0\prime} + \frac{RT}{nF} \ln \frac{\lim_{t\to\infty} C_O^*}{\lim_{t\to\infty} C_R^*} = E^{0\prime} + \frac{RT}{nF} \ln \left[\frac{C_{O,\text{initial}}^* (1-x)}{C_{O,\text{initial}}^* x V_s / V_{Hg}} \right]$$

$$E = E^{0\prime} + \frac{RT}{nF} \ln \left[\frac{V_{Hg}(1-x)}{V_s x} \right]$$

$$E = E^{0\prime} + \frac{RT}{nF} \ln \left[\frac{V_{Hg}}{V_s} \right] + \frac{RT}{nF} \ln \left[\frac{1-x}{x} \right] \tag{3}$$

Solving for x yields the equation for completeness of electrolysis, which is analogous to equation (12.1.5). For V_s and V_{Hg} comparable, as for a mercury pool, x is only slightly shifted, but for a HMDE where $V_{Hg} << V_s$, the impact of the second term in equation (3) can be substantial.

$$\frac{1-x}{x} \left[\frac{V_{Hg}}{V_s} \right] = \exp \left[\frac{nF}{RT} (E - E^{0\prime}) \right]$$

$$x = \frac{1}{1 + \frac{V_s}{V_{Hg}} \exp \left[\frac{nF}{RT} (E - E^{0\prime}) \right]} \tag{4}$$

(c). It is given that $V_s = 200$ cm^3 and $V_{Hg} = 100$ cm^3, so that $V_s/V_{Hg} = 2$. The condition to be met is $x_X > 0.999$ and $x_Y < 0.001$. The applicable potentials are calculated according to equation (3). Conditions are satisfied for X^{2+} when

$$E < -0.45 + \frac{RT}{2F} \ln \left[\frac{1}{2} \right] + \frac{RT}{2F} \ln \left[\frac{1 - 0.999}{0.999} \right]$$

$$< -0.548 \ V$$

Conditions for Y^{2+} require

$$E > -0.70 + \frac{RT}{2F} \ln \left[\frac{1}{2} \right] + \frac{RT}{2F} \ln \left[\frac{1 - 0.001}{0.001} \right]$$

$$> -0.620 \ V$$

Thus, for potentials $-0.548 > E > -0.620$ V, the conditions are satisfied such that $[X^{2+}] \to 0$ and $[Y^{2+}]$ is essentially unchanged.

(d). The electrolysis time needed to reduce the concentration of X^{2+} to $0.001 C_O^*(0)$ is found from equation (12.2.8),

$$t = \frac{-2.303}{p} \log \left[\frac{C_O^*(t)}{C_O^*(0)} \right] \tag{12.2.8}$$

where $p = m_O A / V = 0.01$ cm/s $\times 50$ cm^2/200 cm^3 = 0.0025 s.

$$t = -\frac{2.3}{0.0025 \ \text{s}} \log [0.001] = 2760 \ \text{s} = 46 \ \text{min}$$

Chapter 12   BULK ELECTROLYSIS

Problem 12.6[©]   A.J. Bard developed this problem as a homework assignment during his first year teaching electrochemistry. The principles underlying it are presented in Sections 12.3.2 and 12.4.

This solution is constructed from the indicative curves in Figure 12.9.1, which are approximate with respect to potential. The points to note in those curves are:

- The bromine/bromide couple is reversible with $E^{0\prime}$ near 0.80 V vs. SCE.

- The iodine/iodide couple is reversible with $E^{0\prime}$ near 0.28 V vs. SCE.

- The stannic/stannous couple is shown as irreversible, but with both waves visible. The reduction wave has $E_{1/2}$ at about -0.3 V vs. SCE, and the oxidation wave has $E_{1/2}$ at about 0.1 V vs. SCE.

- The negative background limit is ultimately defined by hydrogen evolution and begins at about -0.55 V, then rises sharply at -0.6 V vs. SCE.

- The positive background limit is defined by oxygen evolution and begins at 1.0 V vs. SCE, then rises sharply at about 1.2 V vs. SCE.

The titration can be conveniently carried out with constant current at a large gold electrode (Table 12.3.1), using apparatus like that in Figure 12.3.5. The sample is said to be roughly equimolar in I_2 and Br_2. Assume that it can be added to the working electrode compartment of a cell containing 0.2 M $NaNO_3$ + 50 mM Sn^{4+} without diluting the electrolyte appreciably, and so that I_2 and Br_2 both fall in the range of 2 mM. The Sn^{4+} is fully oxidized and is stable in the presence of I_2 and Br_2. It is the titrant precursor. The generating cell can be written as:

$$\oplus \text{Au}/NaNO_3\,(0.2\ \text{M})\,/I_2\,(\approx 2\ \text{mM})\,,\ Br_2\,(\approx 2\ \text{mM})\,,\ Sn^{4+}\,(\approx 50\ \text{mM})\,,\ NaNO_3\,(0.2\ \text{M})\,/\text{Au}\ \ominus$$

A separator such as a glass frit or microporous membrane is needed at the junction between the working and counter electrode compartments; junction potentials are not of concern in constant current cells. The working compartment is stirred, and a constant current is applied across the cell with the Au electrode shown on the right being the cathode. The applied current should be mostly supported by the large concentration of Sn^{4+} in the cell. It must not exceed the mass-transfer limited current for reduction of Sn^{4+}. Small portions of the current would also go to the direct reductions of I_2 and Br_2. The electrode reactions at the generating electrode would be

$$Sn^{4+} + 2e \;\rightarrow\; Sn^{2+} \tag{1}$$
$$I_2 + 2e \;\rightarrow\; 2I^- \tag{2}$$
$$Br_2 + 2e \;\rightarrow\; 2Br^- \tag{3}$$

Of the three reductants generated in   reactions (1)–(3), Sn^{2+} is produced at the most negative potential and is the strongest. The order is $Sn^{2+} > I^- > Br^-$; therefore, the following reactions are spontaneous in solution:

$$Sn^{2+} + I_2 \;\rightarrow\; Sn^{4+} + 2I^- \tag{4}$$
$$Sn^{2+} + Br_2 \;\rightarrow\; Sn^{4+} + 2Br \tag{5}$$
$$2I^- + Br_2 \;\rightarrow\; I_2 + 2Br^- \tag{6}$$

The result is that all electrons passed at the generating electrode eventually go to the reduction of Br_2 to Br^- (100% titration efficiency for this reduction) until the Br_2 is exhausted. Next, all electrons passed at generating electrode go eventually to the reduction of I_2 to I^- (100% titration efficiency for this reduction) until the I_2 is exhausted. After that, all current goes to production of Sn^{2+}, which no longer has an oxidized species to reduce and accumulates in solution.

If the composition at the start of the titration is ≈ 2 mM Br_2, ≈ 2 mM I_2, and ≈ 50 mM Sn^{4+}, then the table below gives the composition at later stages. In this table, f is the fraction titrated, where 100% is for full titration of Br_2 and 200% is for full titration of both Br_2 and I_2.

f (%)	$[Br_2]$ (mM)	$[Br^-]$ (mM)	$[I_2]$ (mM)	$[I^-]$ (mM)	$[Sn^{4+}]$ (mM)	$[Sn^{2+}]$ (mM)
0	≈ 2	0	≈ 2	0	≈ 50	0
50	≈ 1	≈ 1	≈ 2	0	≈ 50	0
100	0	≈ 2	≈ 2	0	≈ 50	0
150	0	≈ 2	≈ 1	≈ 1	≈ 50	0
200	0	≈ 2	0	≈ 2	≈ 50	0
250	0	≈ 2	0	≈ 2	≈ 49	≈ 1

(**a**). Suppose that an indicator cell has been added to the system for end-point detection as shown in Figure 12.3.5. In this cell, a small Pt disk indicator electrode is coupled with a reference electrode, say an SCE. If, at the various values of f, the potential of that disk were to be scanned vs. the SCE, a set of $i - E$ curves would be produced. They would resemble those in the figure below.

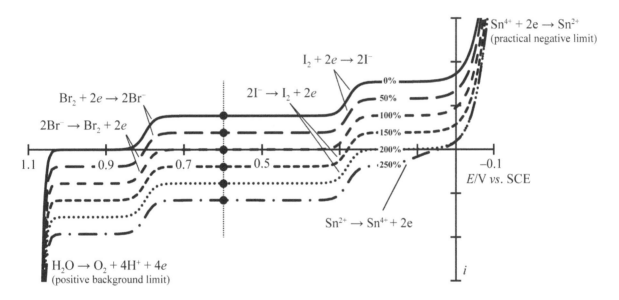

At $f = 0\%$, the indicator electrode shows a reduction wave for Br_2 at 0.8 V and a reduction wave for I_2 at 0.28 V. The positive background limit is the anodic discharge of O_2. The practical negative background limit for this indicator electrode is where Sn^{4+} begins to be reduced. Sn^{4+} is present in the cell at 25 times the concentration of either analyte and can support a large current compared to the voltammetric features of the analytes.

At $f = 50\%$, half of the Br_2 has been converted to Br^-, so there is both a reduction wave for Br_2 and an oxidation wave for Br^-. Because the bromine couple is reversible, they merge into one transition at 0.8V. The reduction wave for I_2 remains.

At $f = 100\%$, all of the Br_2 has been converted to Br^-, so one sees only an anodic wave at 0.8 V for bromide. The reduction wave for I_2 is intact.

At $f = 150\%$, half of the I_2 has been converted to I^-, so one sees a composite anodic and cathodic wave for the iodine couple at 0.28 V. In addition, the anodic wave for Br^- remains.

Beyond $f = 200\%$, there is no more halogen to reduce, so all of the current at the working electrode produces Sn^{2+}, which persists. It is visible voltammetrically as an anodic wave at +0.1 V. In addition, the anodic waves for I^- and Br^- remain.

Amperometry with one indicator electrode is carried out simply by holding that indicator electrode at a fixed potential vs. the reference. A convenient potential for this system is 0.6 V vs. SCE, which is marked in the figure by the vertical dotted line. The current at the indicator electrode is recorded vs. fraction titrated. The currents measured at the values of f represented by the curves are shown as the black circles on the dotted vertical line. The corresponding amperometric plot is shown in the following figure.

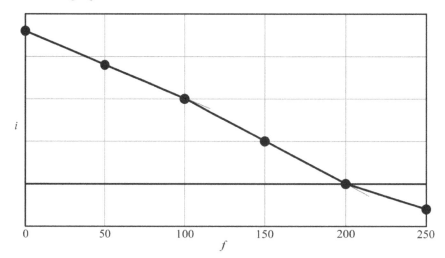

The equivalence point for titration of Br_2 is at 100% and that for titration of I_2 is at 200%. The second is marked by a change in the sign of the current, which makes the end point easier to detect. Usually, there is a slope change at an equivalence point because of differences in the mass-transfer coefficients of species that underlie the change in current before and after the equivalence point (in this case, Br_2 before 100%, I_2 between 100% and 200%, and Sn^{2+} after 200%). The slope change can be used to identify the end point. In a typical titration, there is a continuous record of current at the indicator electrode vs. electrolysis time, so there are far more data than the few points represented here. With a continuous record, a slope change is often clear enough.

(b). Now suppose the reference electrode in the indicator cell of part (a) is replaced with a second Pt indicator electrode identical to the first. Each interacts with the solution according to the $i - E$ curves given above. A voltage differential of 100 mV is imposed between them, so they are always 100 mV apart on the potential axis. The magnitude of the current flowing between the two indicator electrodes is the measurable.

The key to understanding an amperometric titration curve recorded in this mode is to recognize that one of the indicator electrodes must be an anode and the other must be a cathode with currents of equal size always passing at their interfaces. Consequently, they must straddle the zero-current potential, adopting the particular position that allows equal, but opposite, currents at both interfaces. The method, in effect, gauges the approximate slope of the relevant $i - E$ curve at the zero-current potential. If the slope is flat, the electrodes cannot pass much current, but if the slope is steep, they can. These principles work out as follows for the curves relating to this system.

At $f = 0\%$, the zero-current potential is between 0.9 and 1.0 V. The $i - E$ curve is flat in that region, so electrodes 100 mV apart situated on opposite sides of the zero-current potential can pass essentially no current.

At $f = 50\%$, there is a true equilibrium potential near 0.8 V because both Br_2 and Br^- are present. The $i - E$ curve crosses the zero-current axis steeply, so electrodes 100 mV apart straddling this crossing can support substantial current. In fact, the slope of the crossing is at its steepest at this point, because it is right at the inflection point of the composite wave for the bromine/bromide couple. Accordingly, the indicator electrodes will pass a maximum current at this value of f.

At $f = 100\%$, the zero-current potential is between 0.4V and 0.7 V, again in a flat region. The current passing at the indicator electrodes will be small.

At $f = 150\%$, there is a true equilibrium potential near 0.28 V because both I_2 and I^- are present. The $i - E$ curve crosses the zero-current axis steeply, so electrodes 100 mV apart straddling this crossing can support substantial current. As at $f = 50\%$, the crossing is at the inflection point of the relevant wave, so the current will be at a maximum.

At $f = 200\%$, the zero-current potential is between 0.05 V and 0.25 V, yet again in a flat region. Only a very small current can pass between the indicator electrodes.

At $f = 250\%$, there is a true equilibrium potential near 0.05 V supported by Sn^{2+} and Sn^{4+}, which are both present. However, the slope of the $i - E$ curve at the crossing is small because the stannic/stannous couple has slow kinetics. The current at the indicator electrodes will not be zero, but will be much smaller than at $f = 50\%$ or $f = 150\%$.

The titration curve appears as in the figure below, where the behavior between the points has been sketched. In an actual titration carried out in this mode, the whole curve would be recorded continuously. The method is known as "dead-stop" end-point detection, because each end point is marked by a descent to nearly zero current followed by a sharp break in the curve. It is well-suited for use with automatic titration systems.

$$=====$$

Problem 12.7$^{©}$ (a). At the Ag electrode,

$$
\begin{aligned}
Ag + I^- &\rightarrow AgI + e &\text{up to } 100\% \\
Ag &\rightarrow Ag^+ + e &> 100\%
\end{aligned}
$$

For $[I^-] = 1.0 \times 10^{-3}$ mmol/mL in 50 ml, mmol $I^- = 1.0 \times 10^{-3}$ mmol/mL $\times$ 50 ml = 0.050 mmol. From equation (12.2.12),

$$
\begin{aligned}
Q &= nFN_o(0) &(12.2.12)\\
&= 1 \times 96485 \text{ C/mol} \times 5.0 \times 10^{-5} \text{ mol} = 4.8_2 \text{ C}
\end{aligned}
$$

Thus, for

$$
\begin{aligned}
t &= 100 \ s &i = 48.2 \text{ mA} \\
t &= 200 \ s &i = 24.1 \text{ mA, etc.}
\end{aligned}
$$

According to typical applied current ranges and titration times cited on page 504, $i_{app} \approx 48$ mA over 100 s would be suitable.

(b). The $i - E$ curves at a rotating Pt electrode would look as follows:

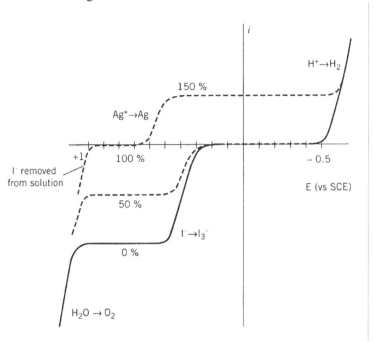

(c). The amperometric titration curves are as shown below for one electrode polarized at -0.3 V vs SCE and for one electrode polarized at +0.4 V vs SCE.

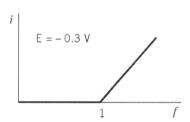
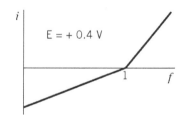

Problem 12.9© Assay of a uranium sample is abbreviated as follows:

$$U_{sample} \xrightarrow{acid} UO_2^{2+} \xrightarrow{Zn(Hg)} U^{3+} \xrightarrow{O_2} U^{4+}$$

(a). After the third step, where $U^{3+} \xrightarrow{O_2} U^{4+}$, the solution contains $[U^{4+}] = 1$ mM. The following species are also added:

$$[Fe^{3+}] = 4 \text{ mM} \qquad [Ce^{3+}] = 50 \text{ mM} \qquad [H_2SO_4] = 1 \text{ M}$$

The goal is to sketch the $i = E$ curve at a Pt rotating disk electrode from -0.2 V to $+1.7$ V vs NHE. The following half-reactions are considered. Note that for reaction (3), the potential is calculated as outlined in problem 2.10.

$$UO_2^{2+} + e \rightleftharpoons UO_2^+ \qquad\qquad E_1^{o'} = 0.05 \text{ V } vs \text{ NHE} \qquad (1)$$
$$UO_2^+ + 4H^+ + e \rightleftharpoons U^{4+} + 2H_2O \qquad E_2^{o'} = 0.62 \text{ V } vs \text{ NHE} \qquad (2)$$
$$UO_2^{2+} + 4H^+ + 2e \rightleftharpoons U^{4+} + 2H_2O \qquad E_3^{o'} = \frac{E_1^{o'} + E_2^{o'}}{2} = 0.335 \text{ V } vs \text{ NHE} \quad (3)$$

$$Fe^{3+} + e \rightleftharpoons Fe^{2+} \qquad\qquad E^{o'} = 0.77 \text{ V } vs \text{ NHE} \qquad (4)$$
$$U^{4+} + 2H_2O \rightleftharpoons UO_2^{2+} + 4H^+ + 2e \qquad -E^{o'} = -0.335 \text{ V } vs \text{ NHE} \quad (5)$$
$$2Fe^{3+} + U^{4+} + 2H_2O \rightleftharpoons 2Fe^{2+} + UO_2^{2+} + 4H^+ \qquad E_{rxn}^{o'} = 0.435 \text{ V } vs \text{ NHE} \quad (6)$$

$$Ce^{4+} + e \rightleftharpoons Ce^{3+} \qquad E^{o'} = 1.44 \text{ V } vs \text{ NHE}$$
$$2H^+ + 2e \rightleftharpoons H_2 \qquad E^{o'} \approx -0.359 \text{ V } vs \text{ NHE}$$

Note that the reduction potential for hydrogen is taken from the potentials shown in Figure 12.9.1. Based on equation (6) and equation (2.1.29),

$$\ln K_{rxn} = \frac{nFE_{rxn}^{o'}}{RT} = \frac{2 \times 96485 \frac{C}{mol} \times 0.435 \text{ V}}{8.31441 \frac{J}{mol\,K} \times 298.15 \text{ K}}$$
$$K_{rxn} = 5.1 \times 10^{14}$$

This demonstrates that the equilibrium of reaction (2) is strongly favored from left to right. Thus, for every mol of U^{4+}, 2 mol of Fe^{3+} are used up, 2 mol of Fe^{2+} are produced, and 1 mol of UO_2^{2+} is produced. After mixing, the key solution components are:

$$[Fe^{3+}] = 4 \text{ mM} - 2 \text{ mM} = 2 \text{ mM} \qquad [U^{4+}] = 0 \qquad [Ce^{3+}] = 50 \text{ mM}$$
$$[Fe^{2+}] = 2 \text{ mM} \qquad\qquad [UO_2^{2+}] = 1 \text{ mM}$$

Chapter 12   BULK ELECTROLYSIS

The $i - E$ curve would look as shown below. Note that the kinetics for UO_2^{2+}/U^{2+} are slow and the reduction wave occurs outside the potential window.

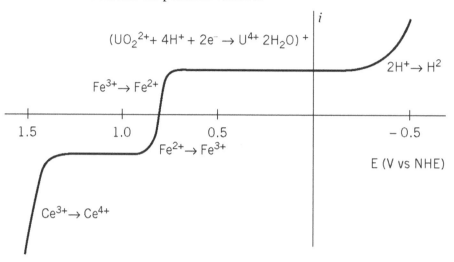

$+ UO_2^{2+}/U^{4+}$ is somewhat irreversible and occurs at more (−) E

(b). The chemistry is outlined in part (a). The coulometric titration will involve the oxidation of $Ce^{3+} \rightarrow Ce^{4+}$ at a Pt electrode followed by the reaction $Ce^{4+} + Fe^{2+} \rightarrow Ce^{3+} + Fe^{2+}$.

(c). For the following stages in the titration, the concentrations are:

Percent Titrated	Iron Species	Cerium Species
0	$\left[Fe^{3+}\right] = \left[Fe^{2+}\right] = 2$ mM	$\left[Ce^{3+}\right] = 50$ mM
50	$\left[Fe^{3+}\right] = 3$ mM; $\left[Fe^{2+}\right] = 1$ mM	$\left[Ce^{4+}\right] = 0$ mM; $\left[Ce^{3+}\right] = 50$ mM
100	$\left[Fe^{3+}\right] = 4$ mM; $\left[Fe^{2+}\right] = 0$ mM	$\left[Ce^{4+}\right] = 0$ mM; $\left[Ce^{3+}\right] = 50$ mM
150	$\left[Fe^{3+}\right] = 4$ mM; $\left[Fe^{2+}\right] = 0$ mM	$\left[Ce^{4+}\right] = 1$ mM; $\left[Ce^{3+}\right] = 49$ mM

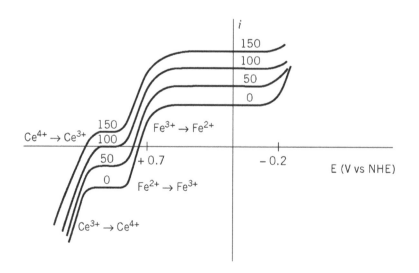

(d). Amperometric titration curves polarized at +0.3 V and +0.9 V:

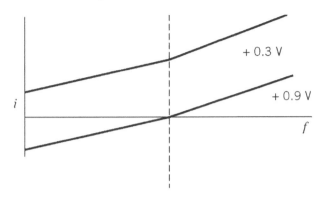

(e). Null current potentiometric responses

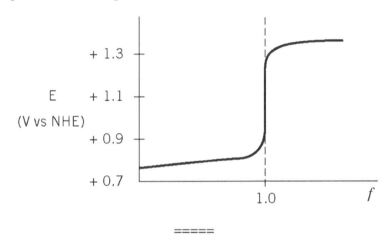

=====

Problem 12.11[©] Consider a solution containing two reducible substances, O_1 and O_2, at concentrations $C^*_{O_1}$ and $C^*_{O_2}$, respectively, where the reversible reduction $O_1 + n_1e \rightleftharpoons R_1$ occurs first, and then, at more negative potentials (e.g., 500 mV separation), the reversible reduction $O_2 + n_2e \rightleftharpoons R_2$. This solution is sandwiched in a thin layer cell of thickness l where there are two working electrodes, such as described in Section 12.6.1. The time of the experiment is much longer than l^2/D, so that mass transfer can be ignored and only electrolysis is important. From Faraday's Law, the moles of O_1 electrolyzed are found.

$$\text{moles of } O_1 \text{ electrolyzed} = \frac{i_1t}{n_1F}$$

From O_1, the concentration of the oxidized species $C_{O_1}(t)$ at time t is expressed. The corresponding concentration of the reduced species is $C_{R_1}(t)$. The volume V is the product of the electrode area A and the cell thickness l.

$$C_{O_1}(t) = C^*_{O_1} - \frac{it}{n_1FV}$$

$$C_{R_1}(t) = \frac{it}{n_1FV}$$

111

Chapter 12   BULK ELECTROLYSIS

Substitution of $C_{O_1}(t)$ and $C_{R_1}(t)$ into the Nernst equation leads to

$$E = E^{0'} + \frac{RT}{n_1 F} \ln \frac{C_{O_1}^* - it/n_1 FV}{it/n_1 FV}$$

The transition time occurs when the numerator equals zero, thus, causing $E \to \infty$. The transition time, τ_1, for the first species, O_1, is then

$$\tau_1 = \frac{n_1 FAlC_{O_1}^*}{i}$$

For chronopotentiometry, the current is specified and the flux of O_1 and O_2 satisfy $i = i_1 + i_2$. The transition time τ is set by the sum of the transition times for the first and second reactions. That is, $\tau = \tau_1 + \tau_2$, where τ_1 corresponds to the transition time for O_1 and τ_2 to the transition time for O_2; τ_2 follows τ_1. Thus, by analog to Section 9.5 for a two-component semi-infinite system,

$$\tau_1 + \tau_2 = \frac{n_1 FAlC_{O_1}^*}{i} + \frac{n_2 FAlC_{O_1}^*}{i} \tag{1}$$

Equation (1) can be recast as

$$i(\tau_1 + \tau_2) = FAl(n_1 C_{O_1}^* + n_2 C_{O_2}^*) \tag{2}$$

Equation (2) demonstrates that transition times depend on the geometry of the thin layer cell and the components in the solution. This equation is compared to equation (9.5.6) for the semi-infinite case for a two-component mixture. Similar to the semi-infinite case, as long as the $i - E$ waves are well-separated, the total current is simply the sum of the individual currents. The differences arise because the thin layer has a boundary set by the thickness l, whereas the semi-infinite solution, has a boundary at ∞. In the semi-infinite case, mass transfer occurs by diffusion and there is a $D^{1/2}$ dependency. In the thin layer cell considered here, under conditions of total electrolysis, there is no mass transfer and thus there is no diffusion coefficient dependency.

=====

Problem 12.13[©]   Metals with either 3 or more electrons in the d orbitals or electrons in the f orbitals, deposit in mercury. J.A. Maxwell and R.P. Graham report all the metals from 3d to 5d for groups 6 (VIA) to 16 (VIB), Cr to Po, deposit in Hg with the exception of W (*Chem. Rev.* **46**, 471 (1950), as shown in Figure 11.3.4 of the second edition). Lanthanides and actinides except Ac are also reported to deposit in Hg. The concentration of metals in seawater are listed below, as reported in the *CRC Handbook of Chemistry and Physics*. Elements are divided into quantitatively deposited, quantitatively separated, and incompletely separated. Where values were readily ascertained, the standard potential for reduction to the metal amalgam are given. For species with mass concentrations similar to or greater than Cu^{2+}, the potentials for reduction of the metal ion to solid metal are reported. x denotes systems where identifying a relevant redox potential is not straightforward but the concentration is sufficiently high that the reaction may also impact the analysis; these species are not considered further.

Concentrations and Some Reduction Potentials vs NHE for Elements in Seawater

	Conc.	$E^{0\,a}$	$E^{0\,b}$		Conc.	$E^{0\,a}$	$E^{0\,b}$
	mg/L	V	V		mg/L	V	V
Quantitatively Deposited				**Quantitatively Separated**			
Cr	3×10^{-4}		-0.90	As	3.7×10^{-3}		x
Fe	2×10^{-3}		-0.44	Se	2×10^{-4}		
Co	2×10^{-5}			Te			
Ni	5.6×10^{-4}		-0.257	Os			
Cu	2.5×10^{-4}	0.345	0.340	Pb	3×10^{-5}	-0.1205	-0.1251
Zn	4.9×10^{-3}	-0.7628	-0.7626				
Ga	3×10^{-5}						
Ge	5×10^{-5}			**Incompletely Separated**			
Mo	1×10^{-2}		-0.200	Mn	2×10^{-4}		-1.18
Tc				Ru	7×10^{-7}		
Rh				Sb	2.4×10^{-4}		x
Pd							
Ag	4×10^{-5}						
Cd	1.1×10^{-4}	-0.3515	-0.4025				
In	2×10^{-2}		-0.3382				
Sn	4×10^{-6}						
Re	4×10^{-6}						
Ir							
Pt							
Au	4×10^{-6}						
Hg	3×10^{-5}						
Tl	1.9×10^{-5}	-0.3338					
Bi	2×10^{-5}						
Po	1.5×10^{-14}						

^{a} E^0 for $M^{n+} + ne \rightleftharpoons Hg(M)$

^{b} E^0 for $M^{n+} + ne \rightleftharpoons M$ where $n = 2$ except for Mo^{3+} and In^{3+}

The species with mass concentrations greater than or approximately equal to that of copper are listed below, as are the concentration (nM) and the approximate potential for each species when the concentration is accounted for according to the Nernst equation for formation of an amalgam ($M^{n+} + ne \rightleftharpoons Hg(M)$). The potentials are calculated using the potential on mercury, where the values are available.

$$E = E^0 + \frac{RT}{nF} \ln \left[M^{n+} \right]$$

All species are treated as dications except the trications of Mo^{3+} and In^{3+}. This analysis is crude because other oxidation states are possible and even probable (especially for Mo, Fe, Cr, and Mn), however, it allows discussion of the important features of the experiment.

	AWT	nM	E
Mo	95.94	104.23	-0.338
Zn	65.39	74.94	-0.974
Fe	55.847	35.81	-0.660
In	114.818	17.42	-0.491
Ni	58.6934	9.54	-0.494
Cr	51.9961	5.77	-1.144
Cu	63.546	**3.93**	0.096
Mn	54.93085	3.64	-1.430
Cd	112.41	0.98	-0.618
Pb	207.2	0.14	-0.291

Clearly, Cu^{2+} has the most positive potential, and it is thus feasible for a potential to be selected where copper alone is plated. The text identifies -0.5 V as the deposition potential, but a somewhat less extreme potential would be better in the context of the approximate values above. A deposition potential of -0.1 V would avoid all the interferences, including Mo^{3+} and Pb^{2+}. Lead engenders the additional complexity of being quantitatively separated but not quantitatively deposited on mercury. When the deposition potential is set to -1.0 V, all of these species except Cr and Mn can be deposited. Thus, it is probable that the additional current reported for -1.0 V is due to the deposition of these additional species along with the Cu.

The concentration of Cu^{2+} in the solution is found by standard addition. The addition of 10^{-7} M Cu^{2+} to the solution generates 0.24 μA or 2.4 A/M. Thus, 0.13 μA corresponds to a concentration of 0.13 μA/2.4 A/M = 54 nM, which is higher than the estimated values in the CRC.

The sensitivity for most polarographic techniques is insufficient to measure concentrations of 50 nM. Polarography is discussed in Chapter 8. On page 361 in the text, the sensitivity of dc polarography is given as $\gtrsim 10^{-5}$ M, with charging current limiting the measurement. Normal pulse polarography has detection limits of 10^{-6} to 10^{-7} M, in part because it discriminates better against charging than does dc polarography. See page 366 in the text. Neither dc nor normal pulse polarography has a low enough detection limit to determine the copper content in seawater. Differential pulse polarography (page 375 in the text) has a detection limit of $\gtrsim 10^{-8}$ M if careful measurements are made. Thus, differential pulse polarography might be used to detect the copper on a HMDE but it would be pushing the limits of detection to make quantitative measurements. Differential pulse could be used to at least qualitatively identify several of the other elements present in seawater.

Problem 12.14[©] **HMDE:** During the 5 min deposition, Pb is collected in the HMDE. As potential is swept positively, Pb is oxidized to Pb^{2+} at the Hg|electrolyte interface and Pb^{2+} diffuses into solution. A peak current of 1 μA at 50 mV/s is reported.

At small electrodes, the peak current i_p is characterized by radial and transient components.

$$i_p(v) = g\left(A_{transient}v^{1/2} + B_{radial}\right) \tag{1}$$

For sufficiently fast v ($A_{transient}v^{1/2} >> B_{radial}$), the observed $i_p(v)$ is set by linear diffusion. For sufficiently slow v ($B_{radial} >> A_{transient}v^{1/2}$), radial diffusion dominates, i_p is v-independent, and voltammograms are sigmoidal.

W.H. Reinmuth (*Anal. Chem.* **33** 185 (1961)) developed equation (1) for a HMDE. For the concentration C_M^* and diffusion coefficient D_M of the metal in the Hg, $g = AD_M^{1/2}C_M^*$. The coefficient $A_{transient} = 2.69 \times 10^5 n^{3/2}$ C(V$^{1/2}$ mol)$^{-1}$ at 25 °C. For sufficiently high v where $B_{radical}$ is negligible, $i_p(v) = gA_{transient}$, which is equation (7.7.21) for linear diffusion to a planar electrode. The radial component, $B_{radial} = -\left(7.25 \times 10^5\right)nD_M^{1/2}r_0^{-1}$, is constant for a given system. (See Section 11.8.2 in *Electrochemical Methods,* 2nd Edition.)

The current scales with g, but g does not impact the transition between linear and radial diffusion. Where $A_{transient}v^{1/2} >> B_{radial}$, linear diffusion dominates and peak currents are observed. For v in V s^{-1}, r_0 in cm, and D_M in cm^2 s^{-1}, where the dimensionless ratio $\left|A_{transient}v^{1/2}/B_{radial}\right| = 0.37r_0\left(nv/D_M\right)^{1/2} \gtrsim 100$, the radial component is negligible and only linear diffusion sets $i_p(v)$. Linear diffusion is favored by larger r_0, higher v, and smaller D_M.

Independent of whether the radial term is negligible or the diffusion is predominantly linear or radial, $i_p(v_2)$ can be determined at a second scan rate v_2 if given given data $i_p(v_1)$ at v_1. Let $b = (v_2/v_1)^{1/2}$. The difference $i_p(v_2) - i_p(v_1)$ eliminates B_{radial} in equation (1).

$$
\begin{aligned}
i_p(v_2) - i_p(v_1) &= gA_{transient}\left[v_2^{1/2} - v_1^{1/2}\right] \\
i_p(v_2) &= i_p(v_1) + gA_{transient}\left[v_2^{1/2} - v_1^{1/2}\right] \\
&= i_p(v_1) + gA_{transient}v_1^{1/2}(b-1) \tag{2}
\end{aligned}
$$

Given $i_p(0.050 \text{ V/s}) = 1$ μA at v_1 of 50 mV/s. At 25 mV/s, $b = (0.025/0.050)^{1/2} = 0.5^{1/2} = 0.71$; at 100 mV/s, $b = (0.100/0.050)^{1/2} = 2^{1/2} = 1.4$. From equation (2),

$$i_p(0.025 \text{ V/s}) = 1\,\mu\text{A} + gA_{transient}(0.050)^{1/2}(0.71 - 1) = 1\,\mu\text{A} - 0.065\,gA_{transient} \tag{3}$$

$$i_p(0.100 \text{ V/s}) = 1\,\mu\text{A} + gA_{transient}(0.050)^{1/2}(1.4 - 1) = 1\,\mu\text{A} + 0.092\,gA_{transient} \tag{4}$$

To find the peak currents, $gA_{transient}$ is needed. If $i_p(v)$ is known experimentally for several scan rates, g and B_{radial} are found from a plot of $i_p(v)$ versus $v^{1/2}$ (equation (1)). Here, only a single datum $i_p(0.05 \text{ V s}^{-1}) = 1$ μA for $n = 2$ is available. If this is recorded under linear diffusion conditions, then $i_p v^{-1/2}$ equals $gA_{transient}$.

To test for linear diffusion, consider common experimental conditions of $r_0 \approx 0.1$ cm and $D_M \approx 10^{-8}$ cm^2 s^{-1}. For $n = 2$ and 0.05 V s^{-1}, $\left|A_{transient}v^{1/2}/B_{radial}\right| = 0.37r_0\left(nv/D_M\right)^{1/2} \approx 120$, which is $\gtrsim 100$. So at 50 mV/s, diffusion is linear. Then, $i_p v^{-1/2} = 4.5 \times 10^{-6}$ A (s/V)$^{1/2} = gA_{transient}$. For $n = 2$, $A_{transient} = 7.61 \times 10^5$ C (V$^{1/2}$ mol)$^{-1}$ and $g = 5.9 \times 10^{-12}$ mol s$^{-1/2}$.

Chapter 12   BULK ELECTROLYSIS

Substitution of $gA_{transient}$ into equations (3) and (4) yields peak currents.

$$
\begin{aligned}
i_p\left(25 \text{ mV/s}\right) &= 1\,\mu\text{A} - 0.065\left(4.5 \times 10^{-6}\right)\text{ A} = 0.71 \times 10^{-6}\text{ A} \\
i_p\left(100 \text{ mV/s}\right) &= 1\,\mu\text{A} + 0.092\left(4.5 \times 10^{-6}\right)\text{ A} = 1.41 \times 10^{-6}\text{ A}
\end{aligned}
$$

At 25 mV/s, $\left|A_{transient}v^{1/2}/B_{radial}\right|$ is ≈ 85, but this is sufficient that $i_p\left(25 \text{ mV/s}\right)$ is governed by linear diffusion and B_{radial} is negligible. For scan rates below 1 mV s^{-1}, B_{radial} is not negligible.

MFE: A mercury film electrode (MFE) behaves as a thin layer where the voltammogram has Gaussian characteristics and no diffusional tailing. Thin layer behavior is favored by low v and thin ℓ so that the concentration profile across the film is flat ($dc(x,t)/dx = 0$ for $0 \leq x \leq \ell$).

To calculate i_p at an MFE, equation (12.6.20) for peak current in a thin layer is used. The volume of the film $V = A\ell$, where ℓ is the thickness of the film.

$$
\left|i_p\right| = \frac{n^2F^2vVC_O^*}{4RT} = \frac{n^2F^2vA\ell C_O^*}{4RT} \tag{12.6.20}
$$

In a thin film, peak current is linearly dependent on scan rate, $\left|i_p\right| \propto v$ or for a given system $\left|i_p\right|/v$ is a constant.

At a mercury film, 25 μA is measured at 50 mV/s. Peak currents for 25 and 100 mV/s are 12.5 μA and 50.0 μA.

$$
\frac{i_p\left(v\right)}{v} = \frac{25\,\mu\text{A}}{50 \text{ mV/s}} = 0.5 \times 10^{-3}\frac{\text{C}}{\text{V}} = 0.05\text{ F} = \frac{12.5\,\mu\text{A}}{25 \text{ mV/s}} = \frac{50\,\mu\text{A}}{100 \text{ mV/s}}
$$

The MFE is on a rotating disk electrode, where $\left|i_p\right|$ is recorded at 2000 rpm and the sweep rate is 50 mV/s. From the Levich equation, equation (10.2.21), current scales with $\omega^{1/2}$ where ω is the rotation rate (s^{-1}).

$$
i_{l,c} = 0.62nFAD_O^{2/3}\omega^{1/2}\nu^{-1/6} \tag{10.2.21}
$$

If ω doubles, currents increase by $\sqrt{2}$ fold. At 4000 rpm in the same system, the current increases.

$$
i^{4000 \text{ rpm}} \approx i^{2000 \text{ rpm}}\left(\frac{4000}{2000}\right)^{1/2} = 25\,\mu\text{A} \times 2^{1/2} = 35.4\,\mu\text{A}
$$

From equation (12.6.20) for at thin film, $\left|i_p\right| \propto \ell$. As ℓ increases, i_p increases proportionally, provided the scan rate is sufficiently slow to maintain thin layer conditions. Once ℓ is too thick or v is too fast, $\left|i_p\right| \propto v^a$, where $1 < a < 0.5$. As a falls below 1, the stripping peaks are broader and less sharp than under thin layer conditions. Once $a = 0.5$, semi-infinite linear diffusion is established in the film, where the concentration at $x = \ell$ is unperturbed from the initial concentration and the voltammogram will exhibit diffusional tailing.

13 ELECTRODE REACTIONS WITH COUPLED HOMOGENEOUS CHEMICAL REACTIONS

Problem 13.1[©] This problem looks at how the cyclic voltammogram for an $E_r C_i E_r$ reaction changes as the time scale of the experiment is varied through the sweep rate. There will be two sets of oxidation-reduction waves: the first corresponding to the reduction $A + e \rightleftharpoons B$, and the second (which occurs after the homogeneous reaction $B \rightarrow C$) corresponding to the reduction $C + e \rightleftharpoons D$.

First of all, referring to Figure 7.2.1, approximately 200 mV is needed to traverse a linear sweep wave. Thus, at a sweep rate of 50 mV/s, the time required to traverse the first peak is about 4 s or 40 half-lives, so that all of B is converted to C through the homogeneous reaction leaving none to be oxidized on the reverse scan. Because all of $B \rightarrow C$, the reduction of $C + e \rightleftharpoons D$ results in a reversible cyclic voltammogram.

At a sweep rate of 1 V/s, the first peak is traversed in approximately 200 ms, corresponding to two half-lives of B. Thus, not all of B is lost to the following homogeneous reaction and a slight peak current is observed on the return sweep. The second wave is still reversible, but the peak heights are smaller than at 50 mV/s.

At a sweep rate of 20 V/s, the time to traverse 0.2 V is 10 ms, which is ten times less than the half-life of B. Thus, only a small amount of B is lost to the following reaction. Thus, the first wave appears almost reversible, and the second wave is reversible but with peak currents less than those seen at 1 V/s.

The following voltammograms were generated using DigiSim 3.0 (Bioanalytical Systems) by M. Rudolph and S.W. Feldberg. The mechanism was specified as

$$A + e \rightleftharpoons B \qquad E^{0\prime} = -0.5 \; V \text{ vs SCE}$$
$$B \xrightarrow{k_f} C \qquad k_f = 6.93 \; s^{-1}$$
$$C + e \rightleftharpoons D \qquad E^{0\prime} = -1.0 \; V \text{ vs SCE}$$

The homogeneous rate constant k_f is estimated from the half life, $\tau_{1/2}$ of 0.1 s for the decay of B to half of its initial concentration B_0. Consider the first order rate expression for decay of B.

$$\frac{dB}{dt} = -k_f B$$

This integrates to

$$\int_{B(0)}^{B(\tau_{1/2})} \frac{dB}{dt} = -k_f \int_0^{\tau_{1/2}} dt = \ln \frac{B_0/2}{B_0} = \ln \frac{1}{2} = -0.693 = -k_f \tau_{1/2}$$

For $\tau_{1/2} = 0.1 \; s$, $k_f = 6.93 \; s^{-1}$

Electrochemical Methods: Fundamentals and Applications, Third Edition, Student Solutions Manual. Cynthia G. Zoski and Johna Leddy. © 2025 John Wiley & Sons Ltd. Published 2025 by John Wiley & Sons Ltd.

Chapter 13 ELECTRODE REACTIONS WITH HOMOGENEOUS REACTIONS

The input parameters consistent with the above mechanism are tabulated as follows.

Estart (V): 0   diffusion: semi-infinite   temperature (K): 298.2
Eswitch (V): -1.2   pre-equilibrium: enabled for all reactions   Ru (Ohms): 0
Eend (V): 0   species parameters: Cdl (F): 0
v (V/s): 0.05   Canal[A] (M/l): 0.001
cycles: 1   Canal[B] (M/l): 0
electrode geometry: planar   Canal[C] (M/l): 0
area (cm2): 1   Canal[D] (M/l): 0

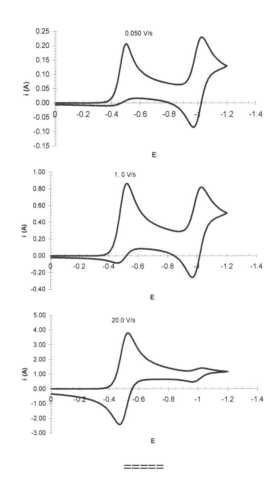

=====

Problem 13.3[©] From the discussion in Section 13.3.1 and Figure 13.3.1, one concludes that the mechanism is E_rC_i.

$$A + e \rightleftharpoons C$$
$$B \xrightarrow{k} C$$

From Chapter 7, a reversible cyclic voltammogram without homogeneous kinetics is characterized by $\Delta E_p = 59/n$ mV at 25 °C and $i_{pr}/i_{pf} = 1$. From the table accompanying the problem, this behavior is seen at sweep rates of 100 V/s and 200 V/s. At sweep rates less than 100 V/s, one sees that the $E_{p/2}$ value shifts in a positive direction from the reversible value at 100 and 200 V/s, i_{pr}/i_{pf} becomes increasingly less than unity. These observations indicate that the immediate product of the electrode reaction, C, undergoes an irreversible reaction. The fact that $i_p/v^{1/2}$ remains nearly constant as the following chemistry is manifested indicates that the following process does not lead to additional electron transfer. Thus, an E_rC_i process is much more likely than E_rC_i', ECE or ECEC.

118

Problem 13.4[©] Cyclic voltammograms for reversible electron transfer, shown in Figure 7.2.5, are described in Section 7.2.2(d). $E_{1/2}$ is located halfway between the forward peak potential E_{pf} and the reverse peak potential E_{pr} irrespective of sweep rate. The half-wave potential is defined as (footnote b under Table 7.2.1)

$$E_{1/2} = E^{0\prime} + \frac{RT}{nF} \ln \left(\frac{D_R}{D_O} \right)^{1/2} \tag{1}$$

Usually D_O and D_R are assumed to be equal so that $E^{0\prime} = E_{1/2}$, thus allowing the standard (or formal) potential to be determined. However, D_O and D_R can be determined from peak currents (equation 7.2.20 for i_{pf}). If D_R/D_O differs from 1, measured D_R/D_O corrects $E_{1/2}$ to yield $E^{0\prime}$. Other assumptions include $n = 1$, semi-infinite linear diffusion, the switching potential is at least $35/n$ mV past E_{pc}, and the absence of homogeneous kinetics.

If the electrode process is EC, then from Figure 13.3.1, one might see a transition from a cyclic voltammogram with only a cathodic peak (for a reduction) to a reversible cyclic voltammogram with both an anodic and cathodic peak as the sweep rate is steadily increased. If this behavior is observed, one can obtain the value of $E_{1/2}$ as the average of the forward and reverse peak potentials in the limit of high sweep rate. At lower sweep rates where the homogeneous decay is evident, one observes a shift in the peak potential, where E_{pf} shifts toward the initial potential of the CV. The shift occurs because of the following reaction consumes the product generated at the electrode. If the reverse peak is missing entirely, it is not practical to use CV to estimate $E_{1/2}$ or E^{0} for a couple. Any estimate made is subject to unknowable error.

If the heterogeneous kinetics are quasireversible, then both E_{pf} and E_{pr} shift out from $E_{1/2}$, such that the peak splitting ΔE_p increases. This is illustrated in curves 3 and 4 of Figure 7.3.2. It has been demonstrated [H. Paul and J. Leddy, *Anal. Chem.* **67** (1995) 1661-1668] that $E_{1/2}$ is approximately (to within 1 mV) midway between the E_{pf} and E_{pr} for $k^0 \gtrsim 0.002$ cm/s and $\Delta E_p \lesssim 145$ mV at 100 mV/s. See also Section 7.3.2(a). As k^0 is increasingly irreversible, the estimate is less accurate.

=====

Problem 13.7[©] Figure 13.3.13 shows cyclic voltammograms for the $C_r E_r$ mechanism where

$$\text{A} \overset{K}{\rightleftharpoons} \text{O}$$
$$\text{O} + \text{e} \rightleftharpoons \text{C}$$

The relevant constants for this problem are as follows: $C_A^* = 1$ mM $= 1 \times 10^{-6}$ mol/cm^3, $D_A = D_O = D_C = 10^{-5}$ cm^2/s, $K = 10^{-3}$, $k_f = 10^{-2}$ s^{-1}, $k_b = 10$ s^{-1}, $T = 25$ °C, and $v = 10$ V/s.

(a). The approximate concentration for species O at the start of the cyclic voltammetric scan is found from equation (13.3.25).

$$
\begin{aligned}
C_O(x,0) &= \frac{C^* K}{K+1} \\
&= \frac{1 \times 10^{-3} \text{ M} \times 10^{-3}}{10^{-3} + 1} = 1 \times 10^{-6} \text{ M} = 1 \times 10^{-9} \frac{\text{mol}}{\text{cm}^3}
\end{aligned}
$$

(b). Assuming that the preceding reaction does not affect the shape of the cyclic voltammogram, the peak current is calculated from equation (6.2.19).

$$i_{pf} = 2.69 \times 10^5 \times 1\ \text{cm}^2 \times \left[10^{-5}\ \frac{\text{cm}^2}{\text{s}}\right]^{1/2} \times 10^{-9}\ \frac{\text{mol}}{\text{cm}^3} \times \left[10\ \frac{\text{V}}{\text{s}}\right]^{1/2} = 2.7 \times 10^{-6}\ \text{A}$$

This peak current is in good agreement with the peak current of $(2.7 - 2.9) \times 10^{-6}$ A observed in Figure 13.3.13 for $v = 10$ V/s.

=====

Problem 13.10© The time reported in τ_{rxn} and τ_{obs} are critical to measuring homogeneous rate constants. Chemical information about the reaction kinetics is embedded in τ_{rxn}. The time scale of the kinetics does not change for a given chemical system. The time scale of the measurement is characterized in τ_{obs}. The range of τ_{obs} is constrained by the measurement protocol, but the time scale for the observation τ_{obs} is controlled by the experimentalist. Best measurements are achieved where τ_{rxn} and τ_{obs} are comparable. If the reaction time τ_{rxn} is substantially faster than the observation time τ_{obs}, the reaction is completed before the measurement time scale probes the reaction time. If τ_{obs} is too fast to allow the kinetics to proceed to a measurable extent, impacts of the chemical reaction will not be observed.

The relationship between measurement time scale and the reaction time scale is defined by λ, the ratio of the measurement time to the reaction time.

$$\lambda = \frac{\tau_{obs}}{\tau_{rxn}}$$

Table 13.4.1 presents expressions for τ_{obs} for the various methods considered in this problem.

Because τ_{rxn} characterizes the rate constants for given reaction conditions (e.g., concentrations, medium, temperature), τ_{rxn} is invariant for measurements on that system. For an EC_i reaction, $\tau_{rxn} = k_f^{-1}$. The time scale of the experiment τ_{obs} varies with the experimental method.

- For $\lambda \lesssim 0.1$ (Zone DO), the experimental method is too fast for the kinetics to proceed during the observation (i.e., τ_{obs} is too short compared to τ_{rxn}), and the measured response will be limited by mass transport of the electroreactant to the electrode surface.

- For $\lambda \gtrsim 10$ (Zone KP), the experimental method is too slow to observe the homogeneous reaction (i.e., τ_{obs} is too long compared to τ_{rxn}), and the measured response will reflect kinetics that are already completely manifested.

- The best range for using a given method to measure kinetic parameters is within an order of magnitude of $\lambda \approx 1$ so that variation in τ_{obs} can match and probe the reaction rate τ_{rxn}. Thus, the optimal range (KO) is where $0.1 \lesssim \lambda \lesssim 10$.

Several measurement methods with different specification of τ_{obs} are considered. The transition from DO through KO to KP is mapped on the EC_i zone diagram.

(a) Rotating ring disk electrode is specified with collection efficiency $i_{ring}/i_{disk} = N = 0.45$.

For an RRDE with rotational frequency ω (s^{-1}), $\tau_{obs} = \omega^{-1}$. Given $\tau_{rxn} = k_f^{-1}$, $\lambda = \tau_{obs}/\tau_{rxn} = k_f/\omega$.

- For diffusion (mass transport limited) conditions (DO), the ring captures the product generated at the disk with the collection efficiency of the device, which is indicated to be 45%; thus,, $i_{ring}/i_{disk} = 0.45$. Product generated at the disk is stable on the measurement time scale τ_{obs}, which is more than an order of magnitude smaller than τ_{rxn} in this zone.

- In zone KP, τ_{obs} is more than an order of magnitude longer than τ_{rxn}, so all product generated at the disk reacts before reaching the ring and $i_{ring} \rightarrow 0$.

- Measurements of k_f are practical only where i_{ring}/i_{disk} falls between 0 and 0.45. In zone KO, the product has time to react but is not completely lost on the time scale of the observation τ_{obs}.

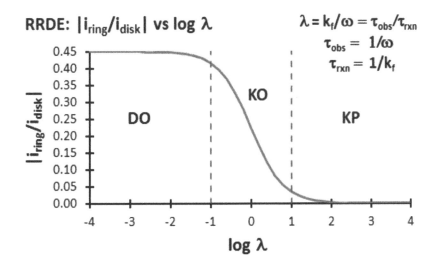

(b) In double-step chronocoulometry, the reactant is first electrolyzed for a time τ (s) and then the potential is reversed and the product is electrolyzed back to reactant. The charge response $Q_d(t)$ is sketched.

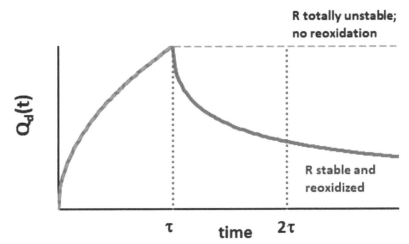

For chronocoulometric measurement time τ, $\tau_{obs} = \tau$. Given $\tau_{rxn} = k_f^{-1}$, $\lambda = \tau_{obs}/\tau_{rxn} = k_f\tau$.

Chapter 13   ELECTRODE REACTIONS WITH HOMOGENEOUS REACTIONS

In one form of the data analysis, charge $Q_d(t)$ is reported at τ and at 2τ. The ratio $Q_d(2\tau)/Q_d(\tau)$ measures loss of product due to chemical reaction.

- Where the product is stable on the time scale of the experiment (zone DO), $Q_d(2\tau)/Q_d(\tau) = 0.414$. The $Q - t$ curve is shown by the solid line.

- Where the time scale of the experiment is long compared to the reaction time (zone KP), all the generated product reacts before being electrolyzed back to reactant. The charge is shown by the dashed line sketched on the chronocoulometric transient. The total charge $Q_d(t)$ is the sum of the charge for the forward and back reactions. Where the product is totally unstable, the charge does not decay from the maximum charge at τ. Then, $Q_d(2\tau)/Q_d(\tau) \longrightarrow 1$.

- Where τ_{obs} is on the same order as τ_{rxn}, $Q_d(2\tau)/Q_d(\tau)$ falls between 0.414 and 1. The larger $Q_d(2\tau)/Q_d(\tau)$, the faster the reaction relative to the measurement.

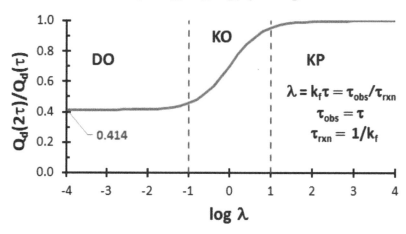

Chronocoulometry: $Q_d(2\tau)/Q_d(\tau)$ vs log λ

$\lambda = k_f \tau = \tau_{obs}/\tau_{rxn}$

$\tau_{obs} = \tau$

$\tau_{rxn} = 1/k_f$

(c) For **square wave voltammetry (SWV)**, the measurement pulse width is t_p (s). With measurement time t_p, $\tau_{obs} = t_p$. Given $\tau_{rxn} = k_f^{-1}$, $\lambda = \tau_{obs}/\tau_{rxn} = k_f t_p$. Impact of the chemical reaction rate is reported as the ratio of the reverse to forward current, $|i_{pr}/i_{pf}|$.

- Where the product is stable on the time scale of the measurement (zone DO), $|i_{pr}/i_{pf}| = 1$.

- Where the product is totally unstable and reacts away on the time scale of the measurement (zone KP), $|i_{pr}| \longrightarrow 0$ and similarly $|i_{pr}/i_{pf}| \longrightarrow 0$.

- Where measurement time scale τ_{obs} is on the order of τ_{rxn} (zone KO), some but not all product reacts on the time scale of the reaction. $0 < |i_{pr}/i_{pf}| < 1$

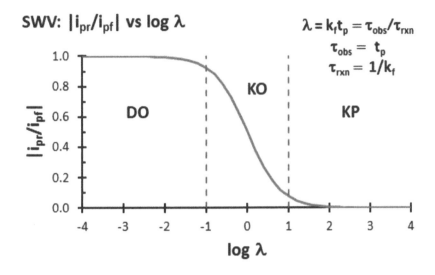

(d) In **chronopotentiometry**, the first current reversal time is t_1 and the reverse transition time is τ_2. Time to first current reversal t_1, $\tau_{obs} = t_1$. Given $\tau_{rxn} = k_f^{-1}$, $\lambda = \tau_{obs}/\tau_{rxn} = k_f t_1$.

- In the mass transport (diffusion) controlled region (zone DO) where product is stable, $\tau_2/t_1 = 1/3$.

- Where product is totally unstable (zone KP), $\tau_2 = 0$ and $\tau_2/t_1 = 0$.

- Where product is partially stable (zone KO), $0 < \tau_2/t_1 < 0.333$.

Chronopotentiometry: τ_2/t_1 vs log λ

Chapter 13   ELECTRODE REACTIONS WITH HOMOGENEOUS REACTIONS

Problem 13.11©   In $\lambda = \tau_{obs}/\tau_{rxn}$, τ_{rxn} is fixed for a given system, but the measurement time τ_{obs} depends on the measurement method and can be varied. Best measurements of kinetic rates are made where $\lambda \approx 1$. Where $\lambda \approx 1$, kinetic decay typically manifests as diminution of the signal by about half.

(a) For cyclic voltammetry (CV) and other voltammetric measurements, $\lambda = 1$ when $\tau_{obs} = \tau_{rxn}$. Table 13.4.1 defines τ_{obs} for various methods.

(b) For CV, $\tau_{obs} = (nfv)^{-1}$, where $f = F/RT$. For EC, $\tau_{rxn} = k_f^{-1}$. Then, $\lambda = \tau_{obs}/\tau_{rxn} = k_f/(nfv)$. For $|i_{pr}/i_{pf}| = 0.5$, $\lambda = 1$ and $nfv = k_f$. For $n = 1$, $k_f = 1000$ s^{-1} and $f = 38.92$ V^{-1} at 25 $^\circ$C, $v = k_f/f = 1000$ s$^{-1}/(38.92V^{-1}) = 25.7$ V/s. This is a fairly fast scan rate for routine work, but far below the limit of the method.

(c) For RRDE, $\tau_{obs} = (\omega)^{-1}$, For EC, $\tau_{rxn} = k_f^{-1}$. Then, $\lambda = \tau_{obs}/\tau_{rxn} = k_f/\omega$. For $\lambda = 1$, $\omega = k_f$. For $k_f = 1000$ s^{-1}, $\omega = 1000$ radian s^{-1}, Rotation rate as frequency is $\omega/(2\pi) = 1000\ s^{-1}/(2\pi) = 159$ s^{-1} or frequency of 9540 rpm. At this rotation rate, i_{ring}/i_{disk} should be about half the collection efficiency N found when the product generated at the disk is stable on the time scale of the experiment. This rotation rate is near the upper limit of the method's range.

(d) For SWV, the measurement pulse width t_p (s) sets $\tau_{obs} = t_p$. Then for SWV, $\lambda = \tau_{obs}/\tau_{rxn} = k_f t_p$. At $|i_{pr}/i_{pf}| = 0.5$, $\lambda = 1$ and $t_p = k_f^{-1}$ For $k_f = 1000$ s^{-1}, $t_p = 1$ ms. This pulse width is short for routine practice, but in mid-range for the method.

14 DOUBLE-LAYER STRUCTURE AND ADSORPTION

Problem 14.2© Consider equation (14.3.10).

$$\left(\frac{d\phi}{dx}\right)^2 = \frac{2kT}{\epsilon\epsilon_0} \sum_j n_j^0 \left[\exp\left(\frac{-z_j e\phi}{kT}\right) - 1\right] \tag{14.3.10}$$

For a symmetrical electrolyte of only two ions where the ions have equal charge, this becomes

$$
\begin{aligned}
\left(\frac{d\phi}{dx}\right)^2 &= \frac{2kTn_j^0}{\epsilon\epsilon_0}\left[\exp\left(\frac{-z_j e\phi}{kT}\right) - 1 + \exp\left(\frac{z_j e\phi}{kT}\right) - 1\right] \\
&= \frac{2kTn_j^0}{\epsilon\epsilon_0}\left[\exp\left(\frac{-z_j e\phi}{kT}\right) + \exp\left(\frac{z_j e\phi}{kT}\right) - 2\right] \\
&= \frac{2kTn_j^0}{\epsilon\epsilon_0}\left[2\cosh\left(\frac{ze\phi}{kT}\right) - 2\right] \\
&= \frac{4kTn_j^0}{\epsilon\epsilon_0}\left[\cosh\left(\frac{ze\phi}{kT}\right) - 1\right]
\end{aligned}
$$

where z is the magnitude of the ionic charge. Take the square root of both sides.

$$\frac{d\phi}{dx} = \pm\sqrt{\frac{4kTn_j^0}{\epsilon\epsilon_0}}\left[\cosh\left(\frac{ze\phi}{kT}\right) - 1\right]^{1/2}$$

But, the half angle formula yields

$$\sqrt{2}\sinh\frac{x}{2} = [\cosh x - 1]^{1/2}$$

Thus, for the negative root, equation (14.3.11) is found.

$$\frac{d\phi}{dx} = -\sqrt{\frac{8kTn_j^0}{\epsilon\epsilon_0}}\sinh\left(\frac{ze\phi}{2kT}\right) \tag{14.3.11}$$

Electrochemical Methods: Fundamentals and Applications, Third Edition, Student Solutions Manual. Cynthia G. Zoski and Johna Leddy.
© 2025 John Wiley & Sons Ltd. Published 2025 by John Wiley & Sons Ltd.

Chapter 14   DOUBLE-LAYER STRUCTURE AND ADSORPTION

Problem 14.4[©]   Consider equation (14.3.29).

$$\sigma^M = \left[8kT\epsilon\epsilon_0 n^0\right]^{1/2} \sinh\left[\frac{ze}{2kT}\left(\phi_0 - \frac{\sigma^M x_2}{\epsilon\epsilon_0}\right)\right] \tag{14.3.29}$$

Expand by noting that $\sinh u = 0.5\left[e^u - e^{-u}\right]$.

$$
\begin{aligned}
\sigma^M &= \frac{\left[8kT\epsilon\epsilon_0 n^0\right]^{1/2}}{2}\left(\exp\left[\frac{ze}{2kT}\left(\phi_0 - \frac{\sigma^M x_2}{\epsilon\epsilon_0}\right)\right] - \exp\left[-\frac{ze}{2kT}\left(\phi_0 - \frac{\sigma^M x_2}{\epsilon\epsilon_0}\right)\right]\right) \\
&= \left[2kT\epsilon\epsilon_0 n^0\right]^{1/2}\left(\exp\left[\frac{ze\phi_0}{2kT}\right]\exp\left[-\frac{ze}{2kT}\frac{\sigma^M x_2}{\epsilon\epsilon_0}\right] - \exp\left[-\frac{ze\phi_0}{2kT}\right]\exp\left[\frac{ze}{2kT}\frac{\sigma^M x_2}{\epsilon\epsilon_0}\right]\right)
\end{aligned}
$$

Differentiate with respect to ϕ_0.

$$
\begin{aligned}
\frac{d\sigma^M}{d\phi_0} &= \left[2kT\epsilon\epsilon_0 n^0\right]^{1/2} \times \\
&\left\{
\begin{aligned}
&\frac{ze}{2kT}\left(\exp\left[\frac{ze\phi_0}{2kT}\right]\exp\left[-\frac{ze}{2kT}\frac{\sigma^M x_2}{\epsilon\epsilon_0}\right] + \exp\left[-\frac{ze\phi_0}{2kT}\right]\exp\left[\frac{ze}{2kT}\frac{\sigma^M x_2}{\epsilon\epsilon_0}\right]\right) \\
&-\frac{ze}{2kT}\frac{x_2}{\epsilon\epsilon_0}\left(\exp\left[\frac{ze\phi_0}{2kT}\right]\exp\left[-\frac{ze}{2kT}\frac{\sigma^M x_2}{\epsilon\epsilon_0}\right] + \exp\left[-\frac{ze\phi_0}{2kT}\right]\exp\left[\frac{ze}{2kT}\frac{\sigma^M x_2}{\epsilon\epsilon_0}\right]\right)\frac{d\sigma^M}{d\phi_0}
\end{aligned}
\right\}
\end{aligned}
$$

Rearrange to yield

$$
\begin{aligned}
&\frac{d\sigma^M}{d\phi_0}\left\{\frac{1}{\left[2kT\epsilon\epsilon_0 n^0\right]^{1/2}} + \frac{ze}{2kT}\frac{x_2}{\epsilon\epsilon_0}\left(
\begin{aligned}
&\exp\left[\frac{ze\phi_0}{2kT}\right]\exp\left[-\frac{ze}{2kT}\frac{\sigma^M x_2}{\epsilon\epsilon_0}\right] \\
&+\exp\left[-\frac{ze\phi_0}{2kT}\right]\exp\left[\frac{ze}{2kT}\frac{\sigma^M x_2}{\epsilon\epsilon_0}\right]
\end{aligned}
\right)\right\} \\
&= \frac{ze}{2kT}\left(\exp\left[\frac{ze\phi_0}{2kT}\right]\exp\left[-\frac{ze}{2kT}\frac{\sigma^M x_2}{\epsilon\epsilon_0}\right] + \exp\left[-\frac{ze\phi_0}{2kT}\right]\exp\left[\frac{ze}{2kT}\frac{\sigma^M x_2}{\epsilon\epsilon_0}\right]\right)
\end{aligned}
$$

Note that $\cosh u = 0.5\left[e^u + e^{-u}\right]$.

$$
\begin{aligned}
&\frac{d\sigma^M}{d\phi_0}\left\{\frac{1}{\left[2kT\epsilon\epsilon_0 n^0\right]^{1/2}} + \frac{ze}{kT}\frac{x_2}{\epsilon\epsilon_0}\cosh\left[\frac{ze}{2kT}\left(\phi_0 - \frac{\sigma^M x_2}{\epsilon\epsilon_0}\right)\right]\right\} \\
&= \frac{ze}{kT}\cosh\left[\frac{ze}{2kT}\left(\phi_0 - \frac{\sigma^M x_2}{\epsilon\epsilon_0}\right)\right]
\end{aligned}
$$

Or,

$$
\begin{aligned}
\frac{d\sigma^M}{d\phi_0} &= \frac{\frac{ze}{kT}\cosh\left[\frac{ze}{2kT}\left(\phi_0 - \frac{\sigma^M x_2}{\epsilon\epsilon_0}\right)\right]}{\frac{1}{\left[2kT\epsilon\epsilon_0 n^0\right]^{1/2}} + \frac{ze}{kT}\frac{x_2}{\epsilon\epsilon_0}\cosh\left[\frac{ze}{2kT}\left(\phi_0 - \frac{\sigma^M x_2}{\epsilon\epsilon_0}\right)\right]} \tag{1} \\
&= \frac{\left[2kT\epsilon\epsilon_0 n^0\right]^{1/2}\frac{ze}{kT}\cosh\left[\frac{ze}{2kT}\left(\phi_0 - \frac{\sigma^M x_2}{\epsilon\epsilon_0}\right)\right]}{1 + \left[2kT\epsilon\epsilon_0 n^0\right]^{1/2}\frac{ze}{kT}\frac{x_2}{\epsilon\epsilon_0}\cosh\left[\frac{ze}{2kT}\left(\phi_0 - \frac{\sigma^M x_2}{\epsilon\epsilon_0}\right)\right]} \\
&= \frac{\left[\frac{2z^2 e^2 \epsilon\epsilon_0 n^0}{kT}\right]^{1/2}\cosh\left[\frac{ze}{2kT}\left(\phi_0 - \frac{\sigma^M x_2}{\epsilon\epsilon_0}\right)\right]}{1 + \left[\frac{2z^2 e^2 \epsilon\epsilon_0 n^0}{kT}\right]^{1/2}\frac{x_2}{\epsilon\epsilon_0}\cosh\left[\frac{ze}{2kT}\left(\phi_0 - \frac{\sigma^M x_2}{\epsilon\epsilon_0}\right)\right]}
\end{aligned}
$$

126

From equations (14.3.27) and (14.3.28),

$$\phi_2 = \phi_0 + \left(\frac{d\phi}{dx}\right)_{x=x_2} x_2$$

$$\left(\frac{d\phi}{dx}\right)_{x=x_2} = -\frac{\sigma^M}{\epsilon\epsilon_0}$$

Then,

$$\phi_2 = \phi_0 - \frac{\sigma^M}{\epsilon\epsilon_0} x_2$$

Equation (1) reduces to equation (14.3.30) where $d\sigma^M/d\phi_0 = C_d$

$$C_d = \frac{d\sigma^M}{d\phi_0} = \frac{\left[\frac{2z^2 e^2 \epsilon\epsilon_0 n^0}{kT}\right]^{1/2} \cosh\left[\frac{ze}{2kT}\phi_2\right]}{1 + \left[\frac{2z^2 e^2 \epsilon\epsilon_0 n^0}{kT}\right]^{1/2} \frac{x_2}{\epsilon\epsilon_0} \cosh\left[\frac{ze}{2kT}\phi_2\right]}$$

The reciprocal yields equation (14.3.31).

$$\frac{1}{C_d} = \frac{x_2}{\epsilon\epsilon_0} + \frac{1}{\left[\frac{2z^2 e^2 \epsilon\epsilon_0 n^0}{kT}\right]^{1/2} \cosh\left[\frac{ze}{2kT}\phi_2\right]} \tag{14.3.31}$$

$=====$

Problem 14.5© In the limit of small ϕ_0, the potential in the diffuse layer decays exponentially.

$$\phi = \phi_0 \exp\left[-\kappa x\right] \tag{14.3.17}$$

From equation (14.3.19), the charge q_s in the diffuse double layer is set by the gradient in potential at the electrode surface.

$$q_s = \varepsilon\varepsilon_0 A \left(\frac{d\phi}{dx}\right)_{x=0} \tag{14.3.19}$$

$$q_s = \varepsilon\varepsilon_0 A \phi_0 (-\kappa) \exp\left[-\kappa x\right]\Big|_{x=0}$$

$$q_s = \frac{-\phi_0 \varepsilon\varepsilon_0 A}{\kappa^{-1}}$$

Because $q_M = -q_s$, the charge on the electrode is $q_M = \phi_0 \varepsilon\varepsilon_0 A/\kappa^{-1}$. The interfacial capacitance of the diffuse layer at small ϕ_0 is given by

$$C = \frac{q_M}{\phi_0} = \frac{\varepsilon\varepsilon_0 A}{\kappa^{-1}} \tag{1}$$

Equation (1) is also recognized as the capacitance of an ideal parallel plate capacitor of thickness κ^{-1}. Thus, the interfacial capacitance is equivalent to that of a parallel plate capacitor of thickness κ^{-1}.

Chapter 14   DOUBLE-LAYER STRUCTURE AND ADSORPTION

At an electrode polarized to potential ϕ_0 relative to bulk solution, potential $\phi(x)$ decays exponentially (top). At $x = \kappa^{-1}$, $\phi(\kappa^{-1}) = 0.37\phi_0$. The linear potential profile of a parallel plate capacitor polarized to ϕ_0 relative to ground is shown (bottom), where the plates are separated by distance κ^{-1}.

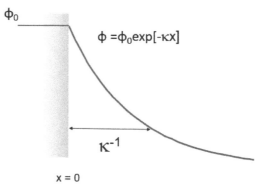

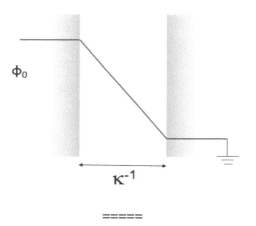

=====

Problem 14.8[©]   Electrocapillary curves are plots of the surface tension of a liquid electrode in contact with a solution. In Figure 14.9.1a, the electrocapillary curve is shown for Na_2SO_4. The potential of zero charge (PZC), where $\sigma^M = \sigma^S = 0$, is found at the maximum of the curve. For Na_2SO_4, the PZC is $\sim -0.8\ V$ vs NCE. In the presence of n-heptanol, the maximum is suppressed because the heptanol is surface active and alters the surface tension. In the absence of n-heptanol, the surface tension of the mercury electrode (γ) is weakened by the charge interactions associated with excess positive and negative charge at the electrode surface. In the presence of n-heptanol, these charge interactions are shielded by the adsorbed alcohol, and the surface tension response is flattened.

The excess charge on the metal, σ^M, is found from the derivative of the plot of γ versus $-E$ according to equation (14.2.1).

$$\sigma^M = -\left(\frac{\partial \gamma}{\partial E_-}\right)_{\mu_{Na_2SO_4},\mu_{Hg}} \tag{14.2.1}$$

In the region where n-heptanol is adsorbed (-0.4 V to -1.6V), the slope of γ versus E is very small, so σ^M is nearly zero.

128

The differential capacitance, C_d, is found from the derivative of the plot of σ^M versus E, according to equation (14.2.2).

$$C_d = \frac{\partial \sigma^M}{\partial E} \tag{14.2.2}$$

Thus, the differential capacitance is the second derivative of the electrocapillary curve with respect to potential.

In Figure 14.9.1b, the differential capacitance curves are shown. The curve for Na_2SO_4 is roughly a gentle parabola, similar to those observed for other electrolytes and modeled by Gouy-Chapman Theory (Figure 14.3.5). For the n-heptanol, the capacitance is roughly invariant between -0.4 and -1.4 V. This is consistent with the adsorbed n-heptanol forming a dielectric, capacitive layer at the interface between the solution and the electrode. Denote the capacitance as C_{hept}. In Grahame's review, he specifies the equivalent circuit for the n-heptanol and Na_2SO_4 system as the resistance of the adsorbed layer in parallel with its capacitance whereas the solution resistance and double layer capacitance are in series. C_{hept} is in series with the double layer capacitance, C_{dl}, which includes the capacitance of the Helmholtz layer and the diffuse layer. For capacitors in series, the total capacitance, C_{total}, is set by the reciprocal sums as $C_{total}^{-1} = C_{dl}^{-1} + C_{hept}^{-1}$. Thus, the smaller capacitance dominates the capacitance of the interface. Between -0.4 and -1.4 V, this is the capacitance of the adsorbed heptanol.

In the presence of n-heptanol, as the potential exceeds the range -0.4 to -1.4 V, the electrode is sufficiently polarized that its charge is compensated by the ions in solution rather than the polar alcohol molecules, and the heptanol is displaced from the electrode surface by the ions. Outside the range -0.4 to -1.4 V, the differential capacitance for the Na_2SO_4 and the Na_2SO_4 with n-heptanol superimpose. The sharp differential capacitance spikes are associated with the desorption of n-heptanol, resulting in a sudden change in the charge in the interfacial region.

$$=====$$

Problem 14.10© For a Langmuirian isotherm where $\theta = \Gamma_j/\Gamma_s$, equation (14.5.8b) applies.

$$\frac{\theta}{1-\theta} = \beta_j a_j^b \tag{14.5.8b}$$

where Γ_s is the saturation coverage of 8×10^{-10} mol/cm^2 and a_j^b is the activity in the bulk. Here, assume the reported β_j includes γ_j/C_j^0 so that the right side of equation (14.8.5b) becomes $\beta_j C_j^b$. If $\theta = 0.5$,

$$\frac{1}{\beta_j} = C_j^b = \frac{1}{5 \times 10^7 \text{ cm}^3/\text{mol}} = 2 \times 10^{-8} \text{ mol/cm}^3$$

The adsorption isotherm is a plot of θ against the solution activity or concentration. From equation (14.5.8b),

$$\theta = \frac{\beta_j C_j^b}{1 + \beta_j C_j^b} \tag{1}$$

For $\beta_j = 5 \times 10^7$ cm^3/mol, this yields the following Langmuirian adsorption isotherm.

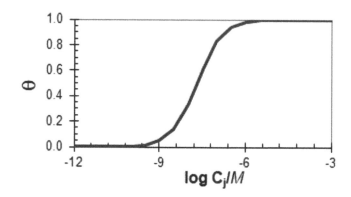

The linearized isotherm applies when, from equation (1), $\theta \sim \beta_j C_j^b$. For this to be correct to within 1%,

$$\frac{\beta_j C_j^b}{1 + \beta_j C_j^b} \geq 0.99 \beta_j C_j^b$$

Or,

$$1 + \beta_j C_j^b \leq \frac{1}{0.99} = 1.01$$
$$\beta_j C_j^b \leq 0.01$$

For $\beta_j = 5 \times 10^7$ cm^3/mol, the linearized version applies for $C_j^b \leq 0.01/\beta_j = 2 \times 10^{-10}$ mol/cm^3. Note that this corresponds to the approximate limit for early linearity in the isotherm above.

$$=====$$

Problem 14.12[©]   First consider the adsorption kinetics for a single species, j, where the change in θ_j (the coverage) is set by the rate of adsorption minus the rate of desorption. The rate of adsorption is set by the rate constant, $k_{a,j}$, the solution concentration, C_j, and the fraction of empty surface sites, $1 - \theta_j$. The rate of desorption is set by the rate constant, $k_{d,j}$, and the coverage. Then,

$$\frac{d\theta_j}{dt} = k_{a,j} C_j (1 - \theta_j) - k_{d,j} \theta_j$$
$$= k_{a,j} C_j - \theta_j [k_{d,j} + k_{a,j} C_j]$$

At steady state, $d\theta_j/dt = 0$. Thus,

$$\frac{d\theta_j}{dt} = 0 = k_{a,j} C_j - \theta_j [k_{d,j} + k_{a,j} C_j]$$
$$\theta_j = \frac{k_{a,j} C_j}{k_{d,j} + k_{a,j} C_j} = \frac{(k_{a,j}/k_{d,j}) C_j}{1 + (k_{a,j}/k_{d,j}) C_j}$$

For $\theta_j = \Gamma_j/\Gamma_s$ and $\beta_j = k_{a,j}/k_{d,j}$, this reduces to equation (14.5.9) for a single species.

$$\Gamma_j = \frac{\Gamma_s \beta_j C_j}{1 + \beta_j C_j} \tag{14.5.9}$$

Now, by analogy consider the competitive adsorption of two species, j and k. The unoccupied surface fraction is $1 - \theta_j - \theta_k$. The steady state rate expressions for the coverage of j and k are defined as follows:

$$\frac{d\theta_j}{dt} = k_{a,j} C_j (1 - \theta_j - \theta_k) - k_{d,j} \theta_j = 0$$

$$\frac{d\theta_k}{dt} = k_{a,k} C_k (1 - \theta_j - \theta_k) - k_{d,k} \theta_k = 0$$

For $\beta_j = k_{a,j}/k_{d,j}$ and $\beta_k = k_{a,k}/k_{d,k}$, the above yield

$$\beta_j C_j (1 - \theta_j - \theta_k) - \theta_j = 0$$
$$-\theta_j \left(\beta_j C_j + 1 \right) - \theta_k \beta_j C_j + \beta_j C_j = 0 \tag{1}$$

$$\beta_k C_k (1 - \theta_j - \theta_k) - \theta_k = 0$$
$$-\theta_j \beta_k C_k - \theta_k \left(\beta_k C_k + 1 \right) + \beta_k C_k = 0 \tag{2}$$

This yields two equations in two unknowns. Equation (1) is rearranged to the following:

$$\theta_k = \frac{\beta_j C_j - \theta_j \left(\beta_j C_j + 1 \right)}{\beta_j C_j}$$

Substitution into equation (2) yields an expression in θ_j.

$$-\theta_j \beta_k C_k - (\beta_k C_k + 1) \left(\frac{\beta_j C_j - \theta_j \left(\beta_j C_j + 1 \right)}{\beta_j C_j} \right) + \beta_k C_k = 0$$

$$\theta_j \left[-\beta_k C_k + \frac{(\beta_k C_k + 1)(\beta_j C_j + 1)}{\beta_j C_j} \right] = -\beta_k C_k + \beta_k C_k + 1$$

$$\theta_j \left[\frac{-\beta_k C_k \beta_j C_j + (\beta_k C_k + 1)(\beta_j C_j + 1)}{\beta_j C_j} \right] = 1$$

$$\theta_j = \frac{\beta_j C_j}{\beta_k C_k + \beta_j C_j + 1}$$

Or, equation (14.5.10a) is found.

$$\Gamma_j = \frac{\Gamma_{j,s} \beta_j C_j}{\beta_k C_k + \beta_j C_j + 1} \tag{14.5.10a}$$

Substitution of θ_j into equation (1) yields equation (14.5.10b).

$$
\begin{aligned}
\theta_k &= \frac{\beta_j C_j - \theta_j \left(\beta_j C_j + 1 \right)}{\beta_j C_j} \\[2mm]
&= \frac{\beta_j C_j - \frac{\beta_j C_j}{\beta_k C_k + \beta_j C_j + 1} \left(\beta_j C_j + 1 \right)}{\beta_j C_j} \\[2mm]
&= 1 - \frac{1}{\beta_k C_k + \beta_j C_j + 1} \left(\beta_j C_j + 1 \right) \\[2mm]
&= \frac{\beta_k C_k + \beta_j C_j + 1 - \left(\beta_j C_j + 1 \right)}{\beta_k C_k + \beta_j C_j + 1} \\[2mm]
&= \frac{\beta_k C_k}{\beta_k C_k + \beta_j C_j + 1}
\end{aligned}
$$

Or, equation (14.5.10b) is found.

$$
\Gamma_k = \frac{\Gamma_{k,s}\beta_k C_k}{\beta_k C_k + \beta_j C_j + 1} \tag{14.5.10b}
$$

=====

Problem 14.13[©]   The Frumkin isotherm accounts for interactions between the adsorbates, either attractive ($g' > 0$) or repulsive ($g' < 0$). Equation (14.5.13) describes the Frumkin isotherm.

$$
\beta_j C_j = \frac{\theta}{1-\theta} \exp\left[-g'\theta \right]
$$

The dimensionless term $\beta_j C_j$ describes the concentration effects. The most direct way to calculate the isotherm is to calculate $\beta_j C_j$ for a range of θ. The isotherm is a plot of θ versus $\beta_j C_j$. The appended spreadsheet shows the responses for g' of 2, 0, and -2. For $g' = 0$, the isotherm is Langmuirian, and on the plot this is the central data set. When $g' = 2$, the interactions are attractive and the adsorbed layer is formed at lower $\beta_j C_j$. Conversely, for $g' = -2$, the interactions are repulsive and higher $\beta_j C_j$ is required to drive monolayer formation.

θ	$\theta/(1-\theta)$	BiCi (g'=0)	BiCi (g'=2)	BiCi (g'=-2)
0.00	0	0	0	0
0.05	0.052632	0.052632	0.047623	0.058167
0.10	0.111111	0.111111	0.09097	0.135711
0.15	0.176471	0.176471	0.130733	0.23821
0.20	0.25	0.25	0.16758	0.372956
0.25	0.333333	0.333333	0.202177	0.549574
0.30	0.428571	0.428571	0.235205	0.780908
0.35	0.538462	0.538462	0.267392	1.084328
0.40	0.666667	0.666667	0.299553	1.483694
0.45	0.818182	0.818182	0.332648	2.012403
0.50	1	1	0.367879	2.718282
0.55	1.222222	1.222222	0.406842	3.671758
0.60	1.5	1.5	0.451791	4.980175
0.65	1.857143	1.857143	0.50613	6.814408
0.70	2.333333	2.333333	0.575393	9.462133
0.75	3	3	0.66939	13.44507
0.80	4	4	0.807586	19.81213
0.85	5.666667	5.666667	1.035207	31.01904
0.90	9	9	1.48769	54.44683
0.95	19	19	2.841804	127.032
0.96	24	24	3.518567	163.703
0.97	32.33333	32.33333	4.646428	224.9996
0.98	49	49	6.902063	347.867
0.99	99	99	13.66885	717.0316
0.997	332.3333	332.3333	45.2471	2440.94

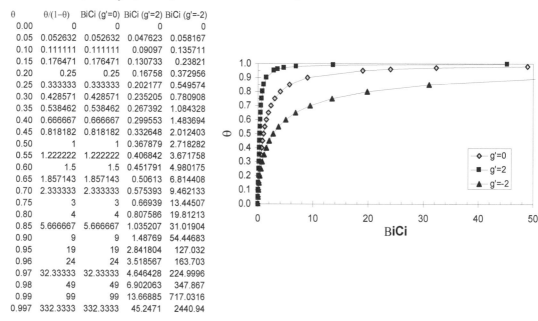

Problem 14.16[©] Outer sphere electron transfer for ferrocene|ferrocenium ($Fc^+ + e \rightleftharpoons Fc$) is shown. The electrode is polarized to potential ϕ^M that establishes formation of the double layer that decays exponentially to the potential of the solution ϕ^S. The potential at the interface drives the electron transfer.

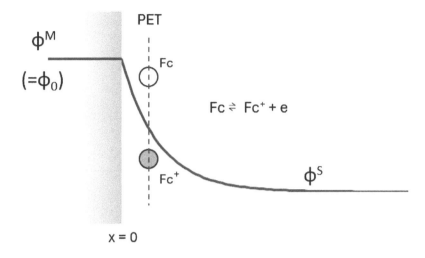

(a) Redox molecules rapidly diffuse and migrate within the electrical double layer (EDL). PET denotes the plane of the electron transfer. For a kinetically fast (reversible) electron transfer under Nernstian conditions, the electrode reaction is at equilibrium at the PET. At equilibrium, the energies of the products (Fc^+ and electrons in the metal) and reactant (Fc) are equal, as described by the electrochemical potentials. From equation (2.2.9), the electrochemical potential for species j in phase β is $\bar{\mu}_j^\beta = \mu_j^{\beta,0} + RT \ln a_j^\beta + z_j F \phi^\beta$.

$$\bar{\mu}_{Fc}^{PET} = \bar{\mu}_{Fc^+}^{PET} + \bar{\mu}_e^M$$
$$\mu_{Fc}^0 + RT \ln a_{Fc}^{PET} = \mu_e^{0,M} - F\phi^M + \mu_{Fc^+}^0 + RT \ln a_{Fc^+}^{PET} + F\phi^{PET}$$

On rearranging,

$$\underbrace{\mu_{Fc}^0 - \mu_e^{0,M} - \mu_{Fc^+}^0}_{\Delta G^0} + F\left(\phi^M - \phi^{PET}\right) = RT \ln \frac{a_{Fc^+}^{PET}}{a_{Fc}^{PET}} \tag{1}$$

The standard free energy ΔG^0 is fixed for a given reaction. As the energy $F\left(\phi^M - \phi^{PET}\right)$ varies, the activities of the electroactive species in the double layer vary. From equation (1), a_{Fc}^{PET} and $a_{Fc^+}^{PET}$ are a function of ϕ^{PET}. Redox activities are a function of the EDL structure at the distance from the electrode where the redox species undergo electron transfer reaction (i.e., at the PET). This will be true for any reaction $O + e \rightleftharpoons R$ because at least either O or R will be a charged species.

(**b**) Equilibrium is established between Fc and Fc^+ at the PET and the bulk solution. The equilibrium is maintained by diffusion and migration of the species between the PET and with the bulk.

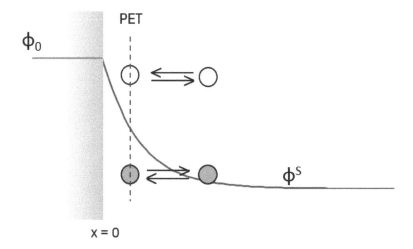

The electrochemical potentials define the equilibrium between the bulk and PET for the neutral Fc and the cationic Fc^+.

$$\bar{\mu}_{Fc}^{PET} = \bar{\mu}_{Fc}^{S}$$
$$\mu_{Fc}^{0} + RT \ln a_{Fc}^{PET} = \mu_{Fc}^{0} + RT \ln a_{Fc}^{S}$$
$$a_{Fc}^{PET} = a_{Fc}^{S} \tag{2}$$

$$\bar{\mu}_{Fc^+}^{PET} = \bar{\mu}_{Fc^+}^{S}$$
$$\mu_{Fc^+}^{0} + RT \ln a_{Fc^+}^{PET} + F\phi^{PET} = \mu_{Fc^+}^{0} + RT \ln a_{Fc^+}^{S} + F\phi^{S}$$
$$RT \ln a_{Fc^+}^{PET} + F\phi^{PET} = RT \ln a_{Fc^+}^{S} + F\phi^{S}$$
$$F\left(\phi^{PET} - \phi^{S}\right) = RT \ln \frac{a_{Fc^+}^{S}}{a_{Fc^+}^{PET}} \tag{3}$$

Elimination of ϕ^{PET} from equation (1) on substitution of equations (2) and (3) yields the relationship between the activities in solution and $\phi^M - \phi^S$.

$$\underbrace{\mu_{Fc}^{0} - \mu_{e}^{0,M} - \mu_{Fc^+}^{0}}_{\Delta G^0} + F\left(\phi^M - \phi^S\right) = RT \ln \frac{a_{Fc^+}^{S}}{a_{Fc}^{S}} \tag{4}$$

Equation (4) relates the activities of Fc and Fc^+ in the bulk solution to the potential difference of the metal relative to the bulk solution. There are no terms for the redox species within the EDL (i.e., no PET terms). The electrode is in equilibrium with Fc and Fc^+ just outside the EDL and this equilibrium is not influenced by the EDL structure.

(c) Equilibrium between the molecules within the EDL region and in the bulk solution is maintained by diffusion and migration. Equilibrium is maintained provided transport is fast as compared to the time required to scan the electrode potential.

For a double layer thickness $d = 10$ nm $= 1 \times 10^{-6}$ cm. ignoring migration, the time required for diffusion across the EDL is estimated from d and $D = 10^{-5}$ cm^2/s.

$$t_{EDL} = \frac{d^2}{2D} = \frac{\left(10^{-6} \text{ cm}\right)^2}{2 \left(10^{-5} \text{ cm}^2/\text{s}\right)} = 5 \times 10^{-8} \text{ s} = 50 \text{ ns}$$

The maximum scan rate is estimated from the characteristic time for sweep voltammetry, $t = RT/Fv = (fv)^{-1}$. The maximum scan rate is estimated from t_{EDL}.

$$v_{\max} = \frac{RT}{Ft_{EDL}} = \frac{0.0256 \text{ V}}{5 \times 10^{-8} \text{ s}} = 5 \times 10^5 \text{ V/s}$$

Equilibrium conditions are maintained by diffusion and the EDL structure does not affect the interfacial nernstian electron transfer provided $v \lesssim 5 \times 10^5$ V/s.

For additional information, see K. J. Levey, M. A. Edwards, H. S. White, and J. V. Macpherson, "Simulation of the cyclic voltammetric response of an outer-sphere redox species with inclusion of electrical double layer structure and ohmic potential drop," *Physical Chemistry Chemical Physics*, **25**(11) 7832–7846 (2023), doi: 10.1039/d3cp00098b.

=====

Problem 14.17[©] The Gauss law (Section 2.2.1) describes the relationship between the charge and the electric field strength vector $\mathcal{E}$ in three dimensions in equation (14.3.18). The dielectric properties of the medium are expressed as the product of the dimensionless dielectric constant of the medium ε and the vacuum electric permittivity ε_0. (See footnote for Section 14.3.1.)

$$q = \varepsilon\varepsilon_0 \oint \vec{\mathcal{E}} \cdot d\vec{S} \qquad \text{(Gauss Law (14.3.18))}$$

In Figure 14.9.2, a Gaussian surface is drawn for a slab that extends infinitely in the yz plane and has a finite thickness Δx in the x direction. The volume contains charge q. In the figure, the electric field is normal to yz plane on each face of the box at x and $x + \Delta x$. The electric field $\mathcal{E}(x)$ points *into* the box at x and $\mathcal{E}(x + \Delta x)$ points *out* of the box at $x + \Delta x$.

The surface integral calculates the electrical flux through all the surfaces of the volume shown in Figure 14.9.2. There is no electrical flux in the infinitely extensive yz plane. Only the electrical flux along the x axis is considered. The surface integral determines the electrical flux from the uniform and constant electrical fields at each x and $x + \Delta x$ through cross sectional area A. Electrical flux out of the box is positive and flux into the box is negative.

$$q = \varepsilon\varepsilon_0 \oint \vec{\mathcal{E}} \cdot d\vec{S} = \varepsilon\varepsilon_0 A \left[\mathcal{E}(x + \Delta x) - \mathcal{E}(x)\right]$$

Chapter 14   DOUBLE-LAYER STRUCTURE AND ADSORPTION

The charge density $\rho\left(x\right)$ is the charge q per volume. The volume is $A\Delta x$.

$$\rho\left(x\right) = \frac{q}{A\Delta x}$$

Then,

$$\rho\left(x\right) = \frac{\varepsilon\varepsilon_0 A\left[\mathcal{E}(x+\Delta x)-\mathcal{E}(x)\right]}{A\Delta x}$$

$$\frac{\rho\left(x\right)}{\varepsilon\varepsilon_0} = \frac{\mathcal{E}(x+\Delta x)-\mathcal{E}(x)}{\Delta x}$$

As $\Delta x \to 0$, the derivative of a continuous function is specified.

$$\lim_{\Delta x\to 0}\frac{\mathcal{E}(x+\Delta x)-\mathcal{E}(x)}{\Delta x} = \frac{d\mathcal{E}(x)}{dx}$$

Then,

$$\frac{\rho\left(x\right)}{\varepsilon\varepsilon_0} = \frac{d\mathcal{E}(x)}{dx} \tag{1}$$

The electric field arises through charge separation. Electrical potential $\phi\left(x,y,z\right)$ is expressed as the work against the electric field strength vector $\vec{\mathcal{E}}(\boldsymbol{x},\boldsymbol{y},\boldsymbol{z})$ in three dimensions in equation (2.2.1). The potential difference for two points in space, $\phi\left(x',y',z'\right)$ and $\phi\left(x,y,z\right)$ is given in equation (2.2.2). In one dimension, the potential difference between $\phi\left(x\right)$ and $\phi\left(x+\Delta x\right)$ is expressed as electric field strength $\mathcal{E}\left(x\right)$.

$$\phi\left(x+\Delta x\right)-\phi\left(x\right) = \int_{x}^{x+\Delta x} -\mathcal{E}(x)dx$$

In derivative form, the electric field is defined by the potential gradient in one dimension.

$$\mathcal{E}(x) = -\frac{\phi\left(x+\Delta x\right)-\phi\left(x\right)}{\Delta x} = -\frac{d\phi}{dx}$$

Substitution of $\mathcal{E}(x)$ defined by potential gradient in equation (1) yields Poisson's equation in one dimension, equation (14.3.5).

$$\frac{\rho\left(x\right)}{\varepsilon\varepsilon_0} = \frac{d\mathcal{E}(x)}{dx} = \frac{d}{dx}\left[-\frac{d\phi}{dx}\right]$$

$$\rho\left(x\right) = -\varepsilon\varepsilon_0\frac{d^2\phi}{dx^2} \qquad \text{(Poisson Equation (14.3.5))}$$

15 INNER-SPHERE ELECTRODE REACTIONS AND ELECTROCATALYSIS

Problem 15.1[C] Tafel's original report of potential E and current density j (A cm^{-2}) for the hydrogen evolution reaction HER are reported by J. Burstein, *Corrosion Science* **47** (2005) 2858 in the preface to a special issue dedicate to Tafel. In the problem statement, the magnitude of the overpotential η is taken as the reported E measured relative to the equilibrium potential E_{eq} to define the magnitude of the overpotential η. In Tafel's data, the potential scale is positive for reduction. In the data analysis below, the sign of η is changed to reflect negative potentials for reduction. HER at Hg is expected to be slow. Interfacial rates of electron transfer measured at $E = E_{eq}$ are reported as exchange current density j_0.

The Tafel equation, equation (15.2.10), $\eta = a + b \log j$ is in linear form. A plot of η (V) versus $\log j$ yields intercept $a = 2.303 RT \left(\alpha n F \right)^{-1} \log j_0$ and slope $b = -2.303 RT \left(\alpha n F \right)^{-1}$. From a plot of η versus $\log j$, $\log j_0 = -a/b$.

(a). Plot of η versus $\log j$ is linear as shown. The Tafel equation is consistent with the data.

η /V	j /A cm^-2	log j
-1.665	0.0004	-3.39794
-1.713	0.001	-3
-1.7465	0.002	-2.69897
-1.7665	0.003	-2.52288
-1.777	0.004	-2.39794
-1.824	0.01	-2
-1.858	0.02	-1.69897
-1.878	0.03	-1.52288
-1.891	0.04	-1.39794
-1.912	0.06	-1.22185
-1.940	0.10	-1
-1.963	0.14	-0.85387
-1.989	0.20	-0.69897

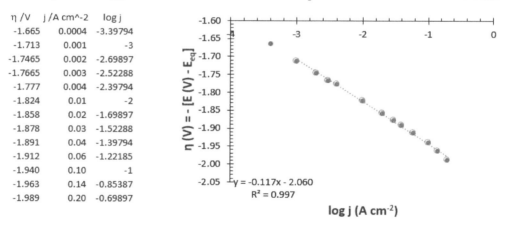

Regression with LINEST in Excel yields $y = - \left(2.060 \pm 0.003 \right) - \left(0.117 \pm 0.002 \right) x$.

(b). The slope of -114 mV is consistent with the -118 mV $= -2.303 RT \left(\alpha n F \right)^{-1}$. At 298.16 K, $-2.303 RT \left(\alpha n F \right)^{-1} = 0.05916\ V/ \left(\alpha n \right)$. For $\alpha n \approx 0.5$, Tafel slope of -118 mV results. For the $n = 1$ and -114 mV, $\alpha \approx 0.5$.

(c). From the regression slope and intercept, $\log j_0 = -a/b = - \left(-2.053 \right) / \left(-0.114 \right) = -18.01$. Then, $j_0 = 10^{-18.0}$ A cm^{-2}, consistent with Figure 15.5.2 where the rate of HER on Hg is among the slowest rates found for sp metals. The value found here from Tafel's data is however, slower than the $\approx 10^{-12}$ A cm^{-2} reported in Figure 15.5.2. HER on Hg is the slowest of any metal. Because of the low HER rate, Hg electrodes provide a larger voltage window for polarography in water.

Electrochemical Methods: Fundamentals and Applications, Third Edition, Student Solutions Manual. Cynthia G. Zoski and Johna Leddy.
© 2025 John Wiley & Sons Ltd. Published 2025 by John Wiley & Sons Ltd.

Chapter 15   INNER-SPHERE REACTIONS AND ELECTROCATALYSIS

Problem 15.3©   The net chlorine evolution reaction (CER) is

$$2Cl^- \rightleftharpoons Cl_2 + 2e$$

is composed of two steps either of which may be rate limiting.

$$Cl^- + L \;\rightleftharpoons\; Cl_{ads} + e \qquad E_V^{0\prime} \tag{15.3.15}$$

$$Cl_{ads} + Cl^- \;\rightleftharpoons\; Cl_2 + L + e \quad E_H^{0\prime} \tag{15.3.16}$$

By analogy to the HER, the subscripts "V" and "H" are used for the first and second steps of CER. Rate determining reaction 15.3.15 for chlorine is analogous to the Volmer step for hydrogen evolution reaction (HER) in equation (15.2.12). Rate determining reaction 15.3.16 for chlorine is analogous to the Heyrovsky step for hydrogen evolution reaction (HER) in equation (15.2.13).

The velocity for the forward (oxidation) reactions 15.3.15 and 15.3.16 v_V and v_H are specified.

$$v_V \;=\; k_V C_{Cl^-}\,(1-\theta)$$

$$v_H \;=\; k_H C_{Cl^-}\,\theta$$

The surface coverage θ is the coverage of Cl_{ads}, the chlorine atom.

(a).   For the rate determining reaction 15.3.15, $i = nFAv_V$. Because the rate for the reaction in equation (15.3.16) is assumed much larger than the reaction in equation (15.3.16), $\theta \to 0$. For Butler Volmer expression for an oxidation, $k_V = k_V^0 \exp\left[(1-\alpha)\,f\left(E - E_V^{0\prime}\right)\right]$. This yields the current expression.

$$
\begin{aligned}
i \;&=\; nFAk_V C_{Cl^-}\\
&=\; nFAk_V^0 C_{Cl^-} \exp\left[(1-\alpha_V)\,f\left(E - E_V^{0\prime}\right)\right]
\end{aligned}
$$

Take the log of both sides and rearrange to the form of the Tafel expression.

$$
\begin{aligned}
\ln\frac{i}{nFA} \;&=\; \ln k_V^0 C_{Cl^-} + (1-\alpha_V)\,f\left(E - E_V^{0\prime}\right)\\[4pt]
E - E_V^{0\prime} \;&=\; \frac{1}{(1-\alpha_V)\,f}\ln\frac{i}{nFA} - \frac{1}{(1-\alpha_V)\,f}\ln k_V^0 C_{Cl^-}
\end{aligned}
$$

Substitution of $\eta = E - E_{eq}$ yields

$$\eta\frac{1}{(1-\alpha_V)\,f}\ln\frac{i}{nFA} - \frac{1}{(1-\alpha_V)\,f}\ln k_V^0 C_{Cl^-} + E_V^{0\prime} - E_{eq}$$

This is analogous to equation (15.2.22) for HER when the Volmer step is rate determining.

The slope of the plot η versus $\log[i/A] = 2.303\left[(1-\alpha_V)\,f\right]^{-1}$. At 25 °C and $\alpha_V = 0.5$, the Tafel slope is 118 mV.

(b). For the rate determining reaction 15.3.16, $i = nFAv_H$. For Butler Volmer expression for an oxidation, $k_H = k_H^0 \exp\left[(1-\alpha) f \left(E - E_V^{0\prime}\right)\right]$. This yields the current expression.

$$
\begin{aligned}
i &= nFAk_H C_{Cl^-} \theta \\
&= nFAk_H^0 C_{Cl^-} \theta \exp\left[(1-\alpha_H) f \left(E - E_H^{0\prime}\right)\right]
\end{aligned}
\tag{1}
$$

When the reaction in equation (15.3.16) is rate determining, θ has a potential dependent, nonzero value. The Nernst equation for the reaction in equation 15.3.15 is used to determine θ as a function of E.

$$
E = E_V^{0\prime} + \frac{RT}{F} \ln\left[\frac{\theta}{(1-\theta)\left(\frac{C_{Cl^-}}{C^0}\right)}\right]
\tag{2}
$$

θ represents the coverage of Cl_{ads}. Rearrangement of (2) yields θ.

$$
\theta = \frac{\frac{C_{Cl^-}}{C^0} \exp\left[f \left(E - E_V^{0\prime}\right)\right]}{1 + \frac{C_{Cl^-}}{C^0} \exp\left[f \left(E - E_V^{0\prime}\right)\right]}
$$

When $E > E_V^{0\prime}$, $\frac{C_{Cl^-}}{C^0} \exp\left[f \left(E - E_V^{0\prime}\right)\right] \gg 1$ and $\theta \to 1$.

When $E < E_V^{0\prime}$, $\frac{C_{Cl^-}}{C^0} \exp\left[f \left(E - E_V^{0\prime}\right)\right] \ll 1$ and

$$
\theta = \frac{C_{Cl^-}}{C^0} \exp\left[f \left(E - E_V^{0\prime}\right)\right]
\tag{3}
$$

(A) At high coverage, $\theta \to 1$ and equation (1) becomes

$$
i = nFAk_H^0 C_{Cl^-} \exp\left[(1-\alpha_H) f \left(E - E_H^{0\prime}\right)\right]
$$

Take the ln of both sides and solve for $E - E_H^{0\prime}$.

$$
\begin{aligned}
\ln\frac{i}{nFA} &= \ln\left(k_H^0 C_{Cl^-}\right) + (1-\alpha_H) f \left(E - E_H^{0\prime}\right) \\
E - E_H^{0\prime} &= \frac{1}{(1-\alpha_H) f} \ln\frac{i}{nFA} - \frac{1}{(1-\alpha_H) f} \ln\left(k_H^0 C_{Cl^-}\right)
\end{aligned}
$$

Or, for $\eta = E - E_{eq}$,

$$
\eta = \frac{1}{(1-\alpha_H) f} \ln\frac{i}{nFA} - \frac{1}{(1-\alpha_H) f} \ln\left(k_H^0 C_{Cl^-}\right) + E_H^{0\prime} - E_{eq}
$$

A plot of η versus $\log[i/A]$ has a slope of $2.303\left[(1-\alpha_H)f\right]^{-1}$. At 25 °C and $\alpha_H = 0.5$, the Tafel slope is 118 mV.

(B) At low coverage, substitution of equation (3) into equation (1) yields the current.

$$
\begin{aligned}
i &= nFAk_H^0 C_{Cl^-} \exp\left[(1-\alpha_H)f\left(E-E_H^{0\prime}\right)\right]\left(\frac{C_{Cl^-}}{C^0}\right)\exp\left[f\left(E-E_V^{0\prime}\right)\right] \\
&= \frac{nFAk_H^0}{C^0}C_{Cl^-}^2 \exp\left[(1-\alpha_H)f\left(E-E_H^{0\prime}\right)\right]\exp\left[f\left(E-E_V^{0\prime}\right)\right]
\end{aligned}
$$

Take the $\ln$ of both sides and solve for $E-E_H^{0\prime}$.

$$
\begin{aligned}
\ln\frac{i}{nFA} &= \ln\left(\frac{k_H^0 C_{Cl^-}^2}{C^0}\right) + (1-\alpha_H)f\left(E-E_H^{0\prime}\right) + f\left(E-E_V^{0\prime}\right) \\
&= \ln\left(\frac{k_H^0 C_{Cl^-}^2}{C^0}\right) + (2-\alpha_H)f\left(E-E_H^{0\prime}\right) - fE_V^{0\prime} + fE_H^{0\prime} \\
E-E_H^{0\prime} &= \frac{\ln\left[\frac{i}{nFA}\right] - \ln\left(\frac{k_H^0 C_{Cl^-}^2}{C^0}\right) + f\left(E_V^{0\prime}-E_H^{0\prime}\right)}{(2-\alpha_H)f}
\end{aligned}
$$

For $\eta = E - E_{eq}$,

$$
\begin{aligned}
\eta &= \frac{\ln\left[\frac{i}{nFA}\right] - \ln\left(\frac{k_H^0 C_{Cl^-}^2}{C^0}\right) + f\left(E_V^{0\prime}-E_H^{0\prime}\right)}{(2-\alpha_H)f} + E_H^{0\prime} - E_{eq} \\
&= \frac{1}{(2-\alpha_H)f}\ln\left[\frac{i}{nFA}\right] - \frac{1}{(2-\alpha_H)f}\ln\left(\frac{k_H^0 C_{Cl^-}^2}{C^0}\right) + \frac{E_V^{0\prime}-E_H^{0\prime}}{(2-\alpha_H)} + E_H^{0\prime} - E_{eq} \\
&= \frac{1}{(2-\alpha_H)f}\ln\left[\frac{i}{nFA}\right] - \frac{1}{(2-\alpha_H)f}\ln\left(\frac{k_H^0 C_{Cl^-}^2}{C^0}\right) + \frac{E_V^{0\prime}}{2-\alpha_H} + E_H^{0\prime}\left(1-\frac{1}{2-\alpha_H}\right) - E_{eq} \\
&= \frac{1}{(2-\alpha_H)f}\ln\left[\frac{i}{nFA}\right] - \frac{1}{(2-\alpha_H)f}\ln\left(\frac{k_H^0 C_{Cl^-}^2}{C^0}\right) + \frac{E_V^{0\prime}}{2-\alpha_H} + \left(\frac{1-\alpha_H}{2-\alpha_H}\right)E_H^{0\prime} - E_{eq}
\end{aligned}
$$

A plot of η versus $\log[i/A]$ has slope $2.303\left[(2-\alpha_H)f\right]^{-1}$. At 25 °C and $\alpha_H = 0.5$, the Tafel slope is 39 mV.

Problem 15.5[©] The exchange current density for the Volmer reaction is expressed in equation (15.5.5)

$$
\begin{aligned}
i_0 &= nFAk_V^0 C_{H^+} \left(\frac{C_{H^+}}{C^0} \right)^{-\alpha_V} (1-\theta)^{1-\alpha_V} \theta^{\alpha_V} \\
&= nFAk_V^0 C_{H^+} \left(\frac{C_{H^+}}{C^0} \right)^{-\alpha_V} (1-\theta) \left[\frac{\theta}{1-\theta} \right]^{\alpha_V}
\end{aligned}
\tag{15.5.5}
$$

The Langmuir isotherm for adsorption of hydrogen is given in equation (15.5.7).

$$
\begin{aligned}
\left[\frac{\theta}{1-\theta} \right]^2 &= \frac{P_{H_2}}{P^0} \exp \left[-\frac{\Delta G^0_{H_{ads}}}{RT} \right] \\
\frac{\theta}{1-\theta} &= \left(\frac{P_{H_2}}{P^0} \right)^{1/2} \exp \left[-\frac{\Delta G^0_{H_{ads}}}{2RT} \right]
\end{aligned}
\tag{15.5.7}
$$

Let $\frac{\theta}{1-\theta} = g = \left(\frac{P_{H_2}}{P^0} \right)^{1/2} \exp \left[-\frac{\Delta G^0_{H_{ads}}}{2RT} \right]$. On rearranging, $\theta = \frac{g}{1+g}$ and $1 - \theta = \frac{1}{1+g}$.

Substitute θ and $1 - \theta$ into equation (15.5.5) to find equation (15.5.8).

$$
\begin{aligned}
i_0 &= nFAk_V^0 C_{H^+} \left(\frac{C_{H^+}}{C^0} \right)^{-\alpha_V} \frac{g^{\alpha_V}}{1+g} \\
&= nFAk_V^0 C_{H^+} \left(\frac{C_{H^+}}{C^0} \right)^{-\alpha_V} \frac{\left(\left(\frac{P_{H_2}}{P^0} \right)^{1/2} \exp \left[-\frac{\Delta G^0_{H_{ads}}}{2RT} \right] \right)^{\alpha_V}}{1 + \left(\frac{P_{H_2}}{P^0} \right)^{1/2} \exp \left[-\frac{\Delta G^0_{H_{ads}}}{2RT} \right]}
\end{aligned}
\tag{15.5.8}
$$

Chapter 15   INNER-SPHERE REACTIONS AND ELECTROCATALYSIS

Problem 15.7[©]   Equation (15.5.8) is given.

$$i_0 = nFAk_v^0 C_{H^+} \left(\frac{C_{H^+}}{C^0}\right)^{-\alpha_V} \frac{\left[\exp\left[-\frac{\Delta G_{H_{ads}}^0}{2RT}\right] \left(\frac{P_{H_2}}{P^0}\right)^{1/2}\right]^{\alpha_V}}{1 + \exp\left[-\frac{\Delta G_{H_{ads}}^0}{2RT}\right] \left(\frac{P_{H_2}}{P^0}\right)^{1/2}}$$

(a). Consider two cases. When $\Delta G_{H_{ads}}^0 >> 0$, then $\lim_{\Delta G_{H_{ads}}^0 \to \infty} \exp\left[-\frac{\Delta G_{H_{ads}}^0}{2RT}\right] \to 0$. The numerator $\to 0$ and the denominator $\to 1$, such that $i_0 \to 0$.
When $\Delta G_{H_{ads}}^0 << 0$, then for the numerator $\lim_{\Delta G_{H_{ads}}^0 \to -\infty} \exp\left[-\frac{\alpha_V \Delta G_{H_{ads}}^0}{2RT}\right] \to \infty$, but for the denominator $\lim_{\Delta G_{H_{ads}}^0 \to -\infty} \exp\left[-\frac{\Delta G_{H_{ads}}^0}{2RT}\right] \to \infty$ faster. Then, $i_0 \to 0$.

(b). For all else fixed and b of the order of 1, i_0 is of the form

$$\frac{i_0}{k} = \frac{b\exp\left[-\alpha_V x\right]}{1 + b\exp\left[-x\right]} \approx \frac{\exp\left[-\alpha_V x\right]}{1 + \exp\left[-x\right]}$$

where $x = \frac{\Delta G_{H_{ads}}^0}{2RT}$. If $x >> 0$, the numerator $\to 0$ and if $x << 0$, then the denominator $\to \infty$. In either case, i_0 is not maximized. To find the maximum, take the derivative with respect to x and set equal to zero.

$$\frac{d\left(i_0/k\right)}{dx} = \left\{\frac{-\alpha_V \exp\left[-\alpha_V x\right]}{1 + \exp\left[-x\right]} - \frac{\exp\left[-\alpha_V x\right]}{\left(1 + \exp\left[-x\right]\right)^2}(-1)\right\} = 0$$

$$\frac{-\alpha_V \exp\left[-\alpha_V x\right]}{1 + \exp\left[-x\right]} = \frac{\exp\left[-\alpha_V x\right]}{\left(1 + \exp\left[-x\right]\right)^2}(-1)$$

$$\alpha_V = \frac{1}{1 + \exp\left[-x\right]}$$

$$\exp[-x] = \frac{1}{\alpha_V} - 1$$

For α_V of the order of 0.5,

$$\exp[-x] \to 1$$

for $x = \frac{\Delta G_{H_{ads}}^0}{2RT} \to 0$. That is, for $\Delta G_{H_{ads}}^0 \to 0$.

16 ELECTROCHEMICAL INSTRUMENTATION

Problem 16.1[©] This problem considers a voltage follower with the input leads reversed so that the feedback loop involves the noninverting input. The circuit diagram is shown below.

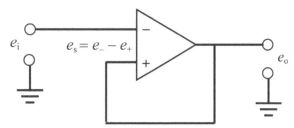

A formula linking the output e_o to the input e_i is derived as follows. First we look at a potential loop between e_o and e_i.

$$-e_o - e_s + e_i = 0$$

Solving for e_o leads to

$$e_o = e_i - e_s \tag{1}$$

From equation (16.1.1)

$$e_s = -\frac{e_o}{A}$$

Substitution into equation (1) leads to

$$e_o = \frac{e_i}{1 - \frac{1}{A}}$$

Because A, the open-loop gain, is very large, the equation simplifies

$$e_o \approx e_i \tag{2}$$

This condition holds for the bandwidth over which the amplification factor remains large enough that $1/A << 1$.

Now suppose that a slight positive fluctuation occurs in e_i vs. ground. Because this change occurs at the inverting input, the output reacts by changing negatively, which shifts the noninverting input negatively, which causes the output to become still more negative, and so on. Thus, the amplifier is driven to its negative limit.

If a negative fluctuation occurs at the input of this circuit in its "equilibrium state," the output reacts by shifting positively, which takes the non-inverting input more positively, which causes the output to move still more positively. The amplifier is driven to its positive limit.

Electrochemical Methods: Fundamentals and Applications, Third Edition, Student Solutions Manual. Cynthia G. Zoski and Johna Leddy.
© 2025 John Wiley & Sons Ltd. Published 2025 by John Wiley & Sons Ltd.

Even though this circuit ostensibly has an equilibrium at $e_o = e_i$, the slightest fluctuation at the input will drive the amplifier to its negative or positive limit. There is no tendency to restore equilibrium. Such instability is characteristic of positive feedback.

The true voltage follower shown in Figure 16.3.1 involves a connection of the output to the inverting input, while e_i is presented at the non-inverting input. The equilibrium condition of the circuit is shown in equation (16.3.3) to be $e_o = e_i$. If a positive fluctuation occurs in e_i *vs.* ground, the amplifier reacts by making a positive change in output, which is fed to the inverting input. This effect counters the fluctuation in e_i and causes the amplifier to tend negatively. Through a continuous process of adjustment, equilibrium is restored. A characteristic of negative feedback is to stabilize the equilibrium condition of a circuit, even if the amplifier tends to overshoot the equilibrium point in the process. In fact, bidirectional overshoot ("ringing") is a common reaction to a sharp step-change in the input voltage.

=====

Problem 16.2[©]   The circuit is formed by combining the adder circuit of Figure 16.2.3 with the integrator circuit of Figure 16.2.4.

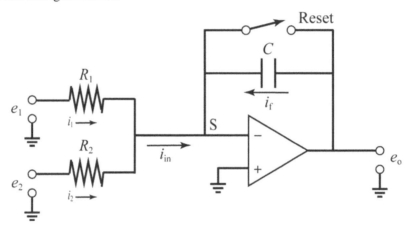

For the adder circuit, as illustrated by equation (16.2.7), the total input current is the sum of the individual input currents.

$$i_{in} = i_1 + i_2$$

From equation (16.2.11) for the integrator,

$$C\frac{de_o}{dt} = -i_{in} \qquad (16.2.11)$$

Thus,

$$C\frac{de_o}{dt} = -(i_1 + i_2)$$

Or, because S is virtual ground,

$$e_o = -\frac{1}{C}\int \left(\frac{e_1}{R_1} + \frac{e_2}{R_2}\right) dt$$

144

Problem 16.4[©]   Consider the modified current follower shown below.

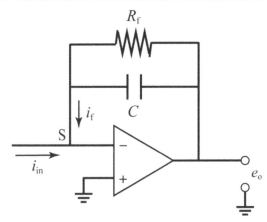

Basically, the capacitor opens a low-impedance feedback path for any high-frequency component in the input signal. Thus, that part of the signal provides little voltage drop across the feedback elements. In consequence, the component of e_o at that frequency is small. Toward any low-frequency component of the input signal, the capacitor is a blocking element. Therefore, the feedback loop acts toward that part of the signal as though the capacitor were not there. In effect, the capacitor helps to filter high-frequency components (which are often noise) from the output of the voltage follower.

The impedance of the parallel resistor and capacitor is developed as outlined in Chapter 11. For elements in parallel, the impedances are summed as reciprocals.

$$\frac{1}{Z} = \frac{1}{R_f} + j\omega C$$

$$Z = \frac{R_f}{1 + j\omega R_f C}$$

By analogy to the text in Section 16.2.1, conservation of charge (i.e., Kirchhoff's laws) dictates that the sum of all the currents into the summing point must be zero. Thus, $i_f = -i_{in}$. The voltage drops around the loop must sum to zero. From Ohm's law,

$$-e_s + e_o + i_f Z = 0$$
$$e_o - e_s = -i_f Z$$

From equation (16.1.1), $e_s = -e_o/A$.

$$e_o\left(1 + \frac{1}{A}\right) = -i_f Z$$

For A very large,

$$e_o \cong -i_f Z$$
$$\cong -i_f \frac{R_f}{1 + j\omega R_f C}$$

If the circuit is subjected to a high ω oscillation or a phase shift, then for sufficiently small C, the capacitor in the feedback loop will filter out the effects of the high frequency oscillation and phase shift. Note that as $\omega \to 0$, the capacitor in the feedback loop has no effect.

In the *IC Op Amp Cookbook*, by W.G. Jung, Prentice Hall, 1997, 3rd Edition, pages 159-160, this circuit is discussed in more detail. For stray capacitance associated with the input, a phase shift can arise. The capacitor in the feedback loop provides a means to compensate for the phase shift. The value of C is found experimentally, with typical values of 3 to 10 μF for $R_f \sim 10$ kΩ.

=====

Problem 16.5© Consider the behavior of the following circuit.

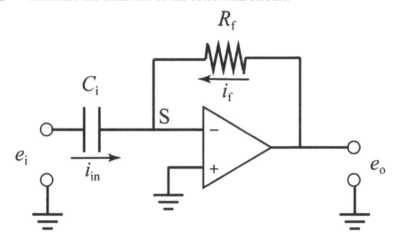

(a) Because the non-inverting input is grounded, the summing point S, is a virtual ground. Thus, e_o is the voltage across R_f, and e_i is the voltage across C_i. Because the input impedance of the amplifier is very high, negligible current flows into the inverting input from the summing point. Accordingly,

$$i_{in} + i_f = 0 \tag{1}$$

The charge q on the capacitor is

$$q = C_i e_i \tag{2}$$

and the time rate of change of the charge is the current, i_{in}.

$$i_{in} = \frac{dq}{dt} = C_i \frac{de_i}{dt} \tag{3}$$

From Ohm's law,

$$i_f = \frac{e_o}{R_f} \tag{4}$$

Substitution of equations (3) and (4) into equation (1) yields

$$C_i \frac{de_i}{dt} + \frac{e_o}{R_f} = 0$$

$$e_o = -R_f C_i \frac{de_i}{dt}$$

The output voltage is proportional to the derivative of the input voltage, with $-R_f C_i$ as the constant of proportionality. The circuit is a differentiator.

(b) The signal-to-noise ratio, S/N, is the amplitude of the information-carrying component divided by the amplitude of the noise component. In this case, one has pure sinusoids for both the information-carrying part and the noise, $e_i = 10 \sin [2\pi (10) t] + 0.1 \sin [2\pi (60) t]$. The amplitude of the signal is 10 and that of the noise is 0.1. The S/N (signal-to-noise) ratio is

$$S/N = \frac{10}{0.1} = 100$$

Even in this problem statement, the word "signal" applies to both the information-carrying component and the combination of that component plus the noise. This usage is rather common, despite the potential confusion. Fortunately, the intended meaning is usually understood from context.

(c) If the composite signal is processed by the differentiator, the result is

$$e_o = -R_f C_i \left(\frac{de_i}{dt} \right) = -R_f C_i \left((10) \, 2\pi \cos [2\pi(10)t] + (0.1) \, 2\pi (60) \cos [2\pi(60)t] \right)$$

The two components are still pure sinusoids, so only the ratio of amplitudes is needed to compute the signal-to-noise ratio. For the information-carrying component, the amplitude of the cosine is $-200\pi R_f C_i$, and for the noise component, the amplitude is $-12\pi R_f C_i$. Thus,

$$S/N = \frac{200}{12} = 17$$

The S/N at the output of the differentiator is 6 times worse than at the input. Differentiation emphasizes the more rapidly changing components in a composite signal, so the signal tends to degrade . Accordingly, differentiators find infrequent use in chemical instrumentation.

(d) For a composite signal is processed by an integrator like that in Figure 16.2.5, the output is

$$e_o = -\frac{1}{R_i C_f} \int e_i dt \qquad (16.2.13)$$

(To preserve consistency with the preceding parts of this problem, the input resistor is labeled as R_i and the feedback capacitor as C_f.)

For the specified composite signal, the output is found.

$$e_o = \frac{-1}{R_i C_f} \int e_i dt = \frac{-1}{R_i C_f} \left[\frac{1 - \cos\left[2\pi(10)t\right]}{2\pi} + \frac{1 - \cos\left[2\pi(60)t\right]}{1200\pi} \right]$$

Then,

$$S/N = \frac{\frac{1}{2}}{\frac{1}{1200}} = 600$$

The S/N at the output of the differentiator is 6 times better than at the input. Integration emphasizes the more slowly changing components in a composite signal, so integrators tends to improve S/N. Integrators are frequently used in chemical instrumentation.

=====

Problem 16.8$^{©}$ (a) In the text, the term "actual potential" means the potential of the working electrode versus the reference without inclusion of any intervening voltage drop in solution. Thus, E is E_{appl} corrected for the ohmic drop. From Section 1.5.4 and (16.7.1),

$$-E = -E_{appl} - iR_u$$

The indicated contact for $-E$ in the dummy cell corresponds to this correction.

(b) An imaginary experiment for measuring actual potential in an operating cell is to move the tip of the reference electrode just outside the outer boundary of the double layer at the working electrode without shielding the working electrode. In that case, iR_u would be negligible and $E_{appl} = E$.

(c) The problem to be treated is a voltage step across a series RC network, which is covered in detail in equations (1.6.6)–(1.6.10). However, the notation is different here. In the present problem, the step causes the voltage at the "Reference" terminal in Figure 16.7.1a to change from 0 V to an arbitrary value, e_{ref}. In the treatment of Chapter 1, the voltage change was from 0 V to an arbitrary value of E. Also, the resistance was labeled R_s in the earlier treatment, rather than R_u here. The relationships $i - t$ and $q - t$ are obtained from equations (1.6.6) and (1.6.10) simply by changing E to e_{ref} and R_s to R_u.

$$i = \frac{e_{ref}}{R_u} \exp\left[-\frac{t}{R_u C_d}\right]$$

$$q = e_{ref} C_d \left(1 - \exp\left[-\frac{t}{R_u C_d}\right]\right)$$

(d) The charge, q, is held on the double-layer capacitance and is the product of the capacitance, C_d, and the voltage across it, which is $-E$. Moreover, $e_{ref} = -E_{appl}$.

$$-EC_d = -E_{appl}C_d \left(1 - \exp\left[-\frac{t}{R_u C_d}\right]\right)$$

$$E = E_{appl} \left(1 - \exp\left[-\frac{t}{R_u C_d}\right]\right)$$

=====

Problem 16.10© If the current follower in the potentiostat of Figure 16.4.5 reaches the output limit (i.e., "goes to the rails"), then the op amp is no longer controlling. This means that the summing point is no longer at virtual ground so the other op amps would no longer control the working electrode against the reference electrode (because they depend on the working electrode is at ground).

If, for example, a step experiment to a negative potential produces a large cathodic current (a positive current into the current follower in the Ohmic convention), the current follower may be driven to its negative limit if $-iR_2$ is more negative than about -13V. If this happens, the CF output can sink the cell current only if the inverting input of CF floats positively away from ground. Thus, the working electrode remains less negative than intended—perhaps substantially less negative— which limits the current to whatever CF can handle in its condition.

The important point is that the actual potential at the working electrode remains less extreme than E_{appl} —more positive when the current is "cathodic," more negative when it is "anodic" (quotes used because the currents need not be faradaic). Consequently, periods at a voltage limit of the current follower are usually not chemically harmful.

Because a voltage-limited current follower impedes the passage of current, it inevitably slows the establishment of the intended potential. In a step experiment, the current follower may be limited in the earliest moments after the step is applied, but as the current demand slackens, the amplifier will come back within its limits, and the potential will reach its intended value.

Chapter 16   ELECTROCHEMICAL INSTRUMENTATION

Problem 16.12[©]   The simple model for the uncompensated resistance in series with the double layer capacitance is a series RC circuit. From Section 1.6.4(a), the current response for the potential step for the series RC circuit is given by equation (1.6.6), where R_s and C_d are the uncompensated solution resistance and the double layer capacitance.

$$i(t) = \frac{E}{R_s} \exp\left(-\frac{t}{R_s C_d}\right) \tag{1.6.6}$$

The data include $E = 0.050$ V, $i(1 \text{ ms}) = 30$ μA, and $i(3 \text{ ms}) = 11$ μA. The ratio of the current responses yields

$$\frac{30 \ \mu\text{A}}{11 \ \mu\text{A}} = 2.73 = \frac{\exp\left(-\frac{10^{-3}}{R_s C_d}\right)}{\exp\left(-\frac{3 \times 10^{-3}}{R_s C_d}\right)} = \exp\left(\frac{2 \times 10^{-3}}{R_s C_d}\right)$$

$$R_s C_d = \frac{2 \times 10^{-3} \text{ s}}{\ln[2.73]} = 2 \text{ ms}$$

Substitution of $R_s C_d$ and E into equation (1.6.6) for $i(1 \text{ ms}) = 30$ μA yields R_s, which in turn yields C_d.

$$\begin{aligned}
R_s &= \frac{E}{i(t)} \exp\left(-\frac{t}{R_s C_d}\right) \\
&= \frac{0.05 \text{ V}}{30 \times 10^{-6} \text{ A}} \exp\left(-\frac{10^{-3}}{2 \times 10^{-3}}\right) \\
&= 1000 \ \Omega
\end{aligned}$$

$$C_d = \frac{1.99 \times 10^{-3} \text{ s}}{1008 \ \Omega} = 2 \ \mu\text{F}$$

Or, for a 0.1 cm^2 electrode, the capacitance is 20 μF/cm^2.

17 ELECTROACTIVE LAYERS AND MODIFIED ELECTRODES

Problem 17.1[©]   The electron transfer for absorbed O_{ads} and R_{ads} is specified in equation (17.2.5).

$$O_{ads} + ne \rightleftharpoons R_{ads} \qquad E^{0\prime}_{ads} \qquad (17.2.5)$$

Consider the case where the system initially contains R_{ads} and the initial potential is well negative of $E^{0\prime}_{ads}$. The initial conditions are only R adsorbed.

$$
\begin{aligned}
\Gamma_R(0) &= \Gamma^* \\
\Gamma_O(0) &= 0
\end{aligned}
$$

At $t = 0$, the potential is swept positive of $E^{0\prime}_{ads}$ at scan rate ν. From equation (17.2.1), the current is set by the flux for reduction of O and equivalently by the flux for the oxidation of R.

$$\frac{i(t)}{nFA} = D_O \left.\frac{\partial C_O(x,t)}{\partial x}\right|_{x=0} - \frac{\partial \Gamma_O(t)}{\partial t} = -\left[D_R \left.\frac{\partial C_R(x,t)}{\partial x}\right|_{x=0} - \frac{\partial \Gamma_R(t)}{\partial t} \right] \qquad (17.2.1)$$

(a). Measurements are made either sufficiently rapidly that the diffusion of O and R in solution do not contribute to measured current or O and R are not initially present in solution. Because current is measured where the diffusion of species in solution does not contribute to the measured current, equation (17.2.6) applies.

$$\frac{i(t)}{nFA} = -\frac{\partial \Gamma_O(t)}{\partial t} = \frac{\partial \Gamma_R(t)}{\partial t} \qquad (17.2.6)$$

Because there is no loss of material from the surface, equation (17.2.7) applies.

$$\Gamma_O(t) + \Gamma_R(t) = \Gamma^* \qquad (17.2.7)$$

For Langmuirian adsorption of O and R as shown in equations (17.2.2) and (17.2.3) with isotherm constants β_O and β_R and saturated surface concentrations $\Gamma_{O,s}$ and $\Gamma_{R,s}$, equation (17.2.8) applies. For $b_O = \beta_O \Gamma_{O,s}$ and $b_R = \beta_R \Gamma_{R,s}$,

$$\frac{\Gamma_O(t)}{\Gamma_R(t)} = \frac{\beta_O \Gamma_{O,s} C_O(0,t)}{\beta_R \Gamma_{R,s} C_R(0,t)} = \frac{b_O C_O(0,t)}{b_R C_R(0,t)} \qquad (17.2.8)$$

The electron transfer is rapid (nernstian), so the ratio $C_O(0,t)/C_R(0,t)$ is set by the Nernst equa-

Electrochemical Methods: Fundamentals and Applications, Third Edition, Student Solutions Manual. Cynthia G. Zoski and Johna Leddy.
© 2025 John Wiley & Sons Ltd. Published 2025 by John Wiley & Sons Ltd.

tion. The formal potential $E^{0\prime}$ for the solution species $O_{soln} + ne \rightleftharpoons R_{soln}$

$$\frac{C_O(0,t)}{C_R(0,t)} = \exp\left[nf\left(E - E^{0\prime}\right)\right] \tag{17.2.9}$$

Substitution of equation (17.2.9) into equation (17.2.8) yields equation (17.2.10). Note from equation (17.2.5), $\Gamma_O(t)/\Gamma_R(t) = \exp\left[nf\left(E - E^{0\prime}_{ads}\right)\right]$.

$$\frac{\Gamma_O(t)}{\Gamma_R(t)} = \frac{b_O}{b_R}\exp\left[nf\left(E - E^{0\prime}\right)\right] = \exp\left[nf\left(E - E^{0\prime}_{ads}\right)\right] \tag{17.2.10}$$

On rearranging, equation (17.2.11) is found.

$$E^{0\prime}_{ads} = E^{0\prime} - \frac{RT}{nF}\ln\frac{b_O}{b_R} \tag{17.2.11}$$

Initially, R is adsorbed, the potential is well negative of $E^{0\prime}_{ads}$; the potential will be swept forward as $E(t) = E_i + \nu t$ for an oxidation. Scan rate $\nu = \partial E/\partial t$. From equation (17.2.6), current is expressed.

$$\frac{i(t)}{nFA} = \frac{\partial\Gamma_R(t)}{\partial t} = \nu\frac{\partial\Gamma_R(t)}{\partial E} \tag{17.2.12}$$

From equations (17.2.7) and (17.2.10),

$$\begin{aligned}
\Gamma_O(t) &= \Gamma_R(t)\exp\left[nf\left(E - E^{0\prime}_{ads}\right)\right] \\
\Gamma^* &= \Gamma_R(t)\left(1 + \exp\left[nf\left(E - E^{0\prime}_{ads}\right)\right]\right) \\
\Gamma_R(t) &= \frac{\Gamma^*}{1 + \exp\left[nf\left(E - E^{0\prime}_{ads}\right)\right]} = \Gamma^*\left(1 + \exp\left[nf\left(E - E^{0\prime}_{ads}\right)\right]\right)^{-1}
\end{aligned}$$

From equation (17.2.12), equation (17.2.14) is found for an oxidation and a positive going potential.

$$\begin{aligned}
\frac{i(t)}{nFA} &= \nu\frac{\partial\Gamma_R(t)}{\partial E} \\
i(t) &= -n^2FAf\nu\Gamma^*\frac{\exp\left[nf\left(E - E^{0\prime}_{ads}\right)\right]}{\left(1 + \exp\left[nf\left(E - E^{0\prime}_{ads}\right)\right]\right)^2} \\
i(t) &= -\frac{n^2F^2A\nu\Gamma^*}{RT}\frac{\exp\left[nf\left(E - E^{0\prime}_{ads}\right)\right]}{\left(1 + \exp\left[nf\left(E - E^{0\prime}_{ads}\right)\right]\right)^2} \tag{17.2.13}
\end{aligned}$$

At the start of the return sweep, only O is adsorbed at coverage Γ^*, the potential is well positive of $E^{0\prime}_{ads}$, and the potential sweep will be negative going to drive the reduction back to R. Analogously to the oxidation, equations (17.2.7) and (17.2.10), $\Gamma_O(t) = \Gamma^* \left[1 + \exp\left[-nf(E - E^{0\prime}_{ads})\right]\right]$. From equation (17.2.6) and $v = \partial E / \partial t$, $i(t)[nFA]^{-1} = -\partial\Gamma_O(t)/\partial t = -v\partial\Gamma_O(t)/\partial E$.

$$i(t) = +\frac{n^2 F^2 A v \Gamma^*}{RT} \frac{\exp\left[nf\left(E - E^{0\prime}_{ads}\right)\right]}{\left(1 + \exp\left[nf\left(E - E^{0\prime}_{ads}\right)\right]\right)^2} \quad \text{for negative going potentials} \quad (17.2.13)$$

Thus, as in Figure 17.2.1, the forward and reverse waves are shaped symmetrically, but the anodic and cathodic currents are of opposite signs.

(b). The current function is of the form $y(x) = B\frac{f(x)}{(1+f(x))^2}$. The maximum or minimum is found when $dy(x)/dx = 0$.

$$
\begin{aligned}
\frac{dy(x)}{dx} &= \frac{d}{dx}\left\{\frac{f(x)}{(1+f(x))^2}\right\} = \frac{d}{dx}\left\{f(x)(1+f(x))^{-2}\right\} \\
&= f'(x)(1+f(x))^{-2} - 2f(x)(1+f(x))^{-3}f'(x) \\
&= f'(x)(1+f(x))^{-2}\left\{1 - 2f(x)(1+f(x))^{-1}\right\} = 0 \\
0 &= 1 - 2f(x)(1+f(x))^{-1} \\
2f(x) &= 1 + f(x) \\
f(x) &= 1 = \exp\left[nf\left(E - E^{0\prime}_{ads}\right)\right] \\
E &= E^{0\prime}_{ads} \text{ at the peak potential}
\end{aligned}
$$

(c). For a nernstian reaction, the peak current occurs at $E = E^{0\prime}_{ads}$, equation (17.2.14).

$$i_p = -\frac{n^2 F^2 A v \Gamma^*}{4RT} \qquad (17.2.14)$$

=====

Problem 17.2[©] The curve in Figure 17.2.5b is almost identical in shape to the theoretical curve in Figure 17.2.5a, consistent with only adsorbed O electroactive. The relationship between peak current, i_p, and surface coverage, Γ^*_O, is given by equation (17.2.26).

$$i_p = \frac{\alpha F^2 A v \Gamma^*_O}{2.718RT} \qquad (17.2.26)$$

To account for n other than 1, the equation is modified as follows, consistent with the usual cluster of nF/RT.

$$i_p = \frac{\alpha n F^2 A v \Gamma^*_O}{2.718RT}$$

It is given that $n = 2$, $A = 0.017$ cm^2, and $v = 0.1$ V/s. From Figure 17.2.5b, $i_p = 2.2 \times 10^{-7}$ A. Assume $\alpha = 0.5$ and $T = 298$ K.

$$i_p = \frac{\alpha n F^2 A v \Gamma_O^*}{2.718RT}$$

$$\Gamma_O^* = \frac{i_p 2.718RT}{\alpha n F^2 A v}$$

$$= \frac{2.2 \times 10^{-7} \text{ A} \times 2.718}{0.5 \times 2 \times 96485 \text{ C/mole} \times 38.92 \text{ V}^{-1} \times 0.017 \text{ cm}^2 \times 0.1 \text{ V/s}}$$

$$= 9.37 \times 10^{-11} \text{ mole/cm}^2$$

In terms of molecules per area,

$$\Gamma_O^* = 9.37 \times 10^{-11} \text{ mole/cm}^2 \times 6.02 \times 10^{23} \text{ molecules/mole}$$

$$= 5.64 \times 10^{13} \text{ molecules/cm}^2$$

This corresponds to 1.77×10^{-14} cm$^2 = 177$ Å^2 or 1.77 nm^2 per molecule of trans-4,4'-dipyridyl-1,2-ethylene.

This is well below a compact monolayer. For example, in Problem 17.7, hydroquinone adsorbs on platinum at 1.6×10^{14} molecules/cm^2 or with a 0.64 nm^2 per molecule. In a close packed monolayer of dipalmitoyl phosphatidyl choline (DPPC) formed by Langmuir Blodgett methods, the footprint of one DPPC molecule is 0.5 nm^2. (*Chem. Rev.* 2022, **122**, 6459–6513)

=====

Problem 17.4[©] This problem relates to the double potential step chronocoulometric study of the induced adsorption of Cd(II) by SCN$^-$ at a HMDE. The data are graphed in Figure 17.3.1. The area of the HMDE is 0.032 cm^2. The potential is stepped from E_i to E_f, and then back to E_i. $\Delta E = E_i - E_f = 0.700$ V

Solution A is composed of 1 mM Cd(II) = 1×10^{-6} mol/cm^3 in 1 M NaNO$_3$. No adsorption occurs in this solution. This is known because the following intercepts are approximately equal.

$$Q_f^0 \cong Q_r^0 \cong 0.55 \ \mu\text{C}$$

Thus, $\Gamma_O = 0$, and equation (17.3.1) $Q_f^0 = nFA\Gamma_O + Q_{dl}$ simplifies to only double layer charging Q_{dl}.

$$Q_{dl} = Q_f^0 = 0.55 \ \mu\text{C}$$

When $\Gamma_O = 0$, $Q_{ads} = nFA\Gamma_O = 0$. Equation (17.3.7)

$$Q = Q_{dl} + Q_{ads} = AC_d(E_i - E_f) + nFA\Gamma_O \tag{17.3.7}$$

can be solved for C_d.

$$C_d = \frac{Q_{dl}}{A\Delta E} = \frac{0.54\ \mu\text{C}}{0.032\ \text{cm}^2 \times 0.7\ \text{V}} = 24\ \mu\text{F/cm}^2$$

Solution B is composed of 1 mM Cd(II) = 1×10^{-6} mol/cm^3 in 0.2 M NaSCN + 0.8 M NaNO$_3$. Adsorption occurs in this solution, as is evidenced by the different values of the intercepts $Q_f^0 = 1.67\ \mu\text{C}$ and $Q_r^0 = 0.86\ \mu\text{C}$. From equation (17.3.6)

$$Q_{dl} = \frac{Q_r^0 - a_o Q_f^0}{1 - a_0} \tag{17.3.6}$$

where a_0 is -0.069, Q_{dl} is found.

$$Q_{dl} = \frac{(0.86 + 0.069 \times 1.67)\mu\text{C}}{1 + 0.069} = 0.91\ \mu\text{C}$$

From equation (17.3.7), capacitance (F cm^{-2}) of the double layer is found.

$$C_d = \frac{Q_{dl}}{A\Delta E} = \frac{0.91\ \mu\text{C}}{0.032\ \text{cm}^2 \times 0.7\ \text{V}} = 41\ \mu\text{F/cm}^2$$

The diffusion coefficient D_O is found from the slope of $Q_f(t)$ vs $t^{1/2}$, where there is no adsorption and the step is to the mass transport limit. From the integral of Cottrell's equation, the slope of a plot of $Q(t)$ versus $t^{1/2}$ yields D_O.

$$Q(t) = 2nFAC_O^* \sqrt{\frac{D_0 t}{\pi}}$$

For Solution A where there is no SCN$^-$, Cd^{2+} diffuses in solution and there is no adsorption. The plot of Q vs $t^{1/2}$ for A$_f$ in Figure 17.3.1 yields a slope of 0.58 μC.

$$slope = S_f^{solutionA} = 2nFAC_O^* D_O^{1/2} \tag{1}$$

where

$$S_f^{solutionA} = 0.58\ \mu\text{C/ms}^{1/2} = \frac{0.58\ \mu\text{C}}{(10^{-3})^{1/2}\text{s}^{1/2}} = 18 \times 10^{-6} \frac{\text{C}}{\text{s}^{1/2}}$$

Solving equation (1) for D_O,

$$D_O = \left(\frac{S_f^{solutionA}}{2nFAC_O^*}\right)^2 = \left(\frac{18 \times 10^{-6} \frac{\text{C}}{\text{s}^{1/2}}}{2 \times 2 \times 96485 \frac{\text{C}}{\text{mol}} \times 0.032\ \text{cm}^2 \times 10^{-6} \frac{\text{mol}}{\text{cm}^3}}\right)^2 = 6.6 \times 10^{-6} \frac{\text{cm}^2}{\text{s}}$$

Surface coverage in the presence of thiocyanate is found from equation (17.3.1).

$$Q_f^0 = nFA\Gamma_O + Q_{dl} \tag{17.3.1}$$

Then, in the presence of thiocyanate, the surface coverage is Γ_O.

$$\Gamma_O = \frac{Q_f^0 - Q_{dl}}{nFA} = \frac{(1.67 - 0.91) \times 10^{-6}\ \text{C}}{2 \times 96485 \frac{\text{C}}{\text{mol}} \times 0.032\ \text{cm}^2} = 1.2 \times 10^{-10} \frac{\text{mol}}{\text{cm}^2}$$

Problem 17.7[©]   **(a).** The moles of hydroquinone HQ in the cell are determined from the concentration and cell volume.

$$\text{moles of HQ} = 0.1 \times 10^{-6} \text{ moles/cm}^3 \times 1.2 \text{ cm}^2 \times 4.0 \times 10^{-3} \text{ cm}$$
$$= 4.8 \times 10^{-10} \text{ moles}$$

The cell volume is 4.8×10^{-3} cm^3. It is given that after filling the cell the first time, electrolysis of the solution solubilized species yields 32 μC. Faraday's law allows the calculation of the moles of material HQ in solution. Under the experimental conditions, only the solution species can be electrolyzed. $n = 2$.

$$\frac{Q}{nF} = \text{moles}$$
$$= \frac{32 \times 10^{-6} \text{ C}}{2 \times 96485 \text{ C/mole}} = 1.7 \times 10^{-10} \text{ moles}$$

Thus, from the initial number of moles in the cell (4.8×10^{-10}) and the moles remaining in solution after adsorption (1.7×10^{-10}), the moles of adsorbed HQ are found: $4.8 \times 10^{-10} - 1.7 \times 10^{-10} = 3.1 \times 10^{-10}$. The cell is emptied and supplied with a fresh aliquot of hydroquinone solution. The electrolysis of the solution species requires 96 μC, which from Faraday's law, corresponds to 5.0×10^{-10} moles or essentially, all of the hydroquinone provided by the fresh aliquot. Thus, all the HQ adsorbed from the first aliquot and prevents adsorption of HQ from the second aliquot. This is consistent with a monolayer formed of closely packed HQ molecules. The area per molecule is then calculated as follows:

$$\Gamma_O = \frac{3.1 \times 10^{-10} \text{ moles}}{1.2 \text{ cm}^2} \times 6.02 \times 10^{23} \text{ molecules/mole}$$
$$= 1.56 \times 10^{14} \text{ molecules/cm}^2$$

The cross sectional area per molecule, σ, is found from the reciprocal.

$$\sigma = \frac{1}{\Gamma_O}$$
$$= \frac{\text{cm}^2}{1.56 \times 10^{14} \text{ molecules}} \times \left(\frac{10^7 \text{ nm}}{\text{cm}} \right)^2$$
$$= 0.64 \text{ nm}^2/\text{molecule}$$

(b). The two probable orientations for adsorbed hydroquinone are flat and edge on. The area of 0.64 nm^2/molecule corresponds to about 0.8 nm between HQ centers. HQ adsorption parallel to the surface is most likely. Adsorption where HQ lies flat is consistent with strong interactions of the aromatic ring with the metal surface.

18 SCANNING ELECTROCHEMICAL MICROSCOPY

Problem 18.1[©] **(a).** The system described is for a feedback configuration and the current response is characterized by equation (18.2.2) for a conductive substrate.

$$\frac{i_T^c}{i_{T,\infty}} = I_T(L) = 0.68 + \frac{0.78377}{L} + 0.3315 \exp\left[-\frac{1.0672}{L}\right] \tag{18.2.2}$$

where $L = d/a$ and $i_{T,\infty} = 4nFD_OC_O^*a$, as shown in equation (18.1.1). The tip radius is a and the distance from the surface is d. It is given that $a = 5.0 \times 10^{-4}$ cm, $C_O^* = 5.0 \times 10^{-6}$ mol/cm^3, $D_O = 5.0 \times 10^{-6}$ cm^2/s, and $i_T/i_{T,\infty} = 2.5$. Equation (18.2.2) is nonlinear and must be fit either from a working curve or by successive approximation. A working curve that incorporates a successive approximation for this case is shown in a spreadsheet. From the curve, $L = 0.438 = d/a$. Thus, $d = 0.438 \times 5.0 \times 10^{-4}$ cm $= 2.19 \times 10^{-4}$ cm $= 2.19$ μm.

L	IT(L)	L	IT(L)
0.1	8.518	1.6	1.340
0.2	4.600	1.8	1.299
0.3	3.302	2	1.266
0.4	2.662	2.2	1.240
0.438	**2.498**	2.4	1.219
0.5	2.287	2.6	1.201
0.6	2.042	2.8	1.186
0.7	1.872	3	1.174
0.8	1.747	3.2	1.162
0.9	1.652	3.4	1.153
1	1.578	3.5	1.148
1.2	1.469	4	1.130
1.4	1.395	4.5	1.116
1.5	1.365255	5	1.105

(b). From equation (18.1.1), $i_{T,\infty} = 4nFD_OC_O^*a$. Let $n = 1$.

$$
\begin{aligned}
i_{T,\infty} &= 4nFD_OC_O^*a \\
&= 4 \times 96,485 \text{ C/mol} \times 5.0 \times 10^{-6} \text{ cm}^2/\text{s} \times 5.0 \times 10^{-6} \text{ mol/cm}^3 \times 5.0 \times 10^{-4} \text{ cm} \\
&= 4.82 \text{ nA.}
\end{aligned} \tag{18.1.1}
$$

(c). Equation (18.2.1) applies to an insulating substrate, $I_T^i(L)$, where the dimensionless geometric factor $RG = r_g/a = 10$. The radius of the insulating shroud is r_g.

$$\frac{i_T}{i_{T,\infty}} = I_T^i(L) = \left[0.4572 + \frac{1.4604}{L} + 0.4313 \exp\left(-\frac{2.3507}{L}\right)\right]^{-1} - 0.14544\frac{L}{5.5769 + L} \tag{18.2.1}$$

For $L = 0.438$, $I_T^i(L) = 0.253$. Here, $I_T^i(L) < 1$, consistent with the insulating substrate restricting access to the tip and decreasing current from that expected for a microdisk in bulk solution.

Electrochemical Methods: Fundamentals and Applications, Third Edition, Student Solutions Manual. Cynthia G. Zoski and Johna Leddy.
© 2025 John Wiley & Sons Ltd. Published 2025 by John Wiley & Sons Ltd.

Problem 18.2[©]   The flux lines for diffusion are qualitatively sketched for a disk SECM tip approaches an insulating substrate where distance of the tip from the surface varies as $L = 100$ (far), 10, and 1 (near contact). The second parameter that impacts flux is the thickness of the insulating shroud around the tip. Flux from the back side of the tip around the shroud is characterized by RG. Both L and RG are dimensionless and relative to the disk radius a.

$$L = \frac{d}{a} = \frac{\text{tip to substrate distance}}{\text{disk radius}} \qquad RG = \frac{r_g}{a} = \frac{\text{sheath radius}}{\text{disk radius}}$$

(a) Diffusional flux lines are shown for the insulating shroud radius tenfold larger than the tip radius, $RG = 10$. Diffusion from behind the plane of the tip is minimal.

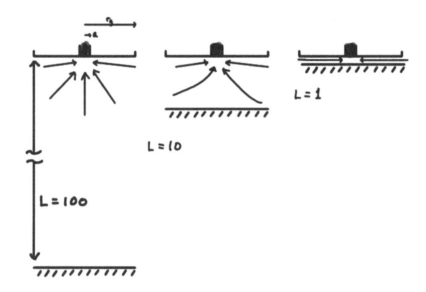

(b) For the insulating shroud radius equal to the tip radius, $RG = 1$, diffusion from behind the plane of the tip and around the edges of the shroud contributes substantially to flux of the redox probe to the tip electrode.

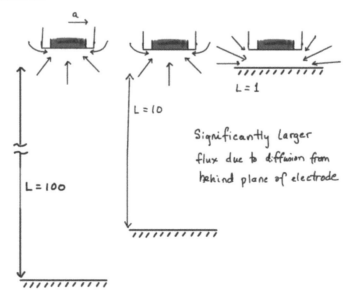

Problem 18.3© Voltammograms of i_T versus E_S are calculated for values of the standard heterogeneous electron transfer rate k^0 of 10^{-2}, 10^{-4}, and 10^{-6} cm/s. The reaction at the tip is $O + e \rightarrow R$ where $L = 0.1$, $a = 10$ μm, $RG = 10$, $D_0 = D_R = 10^{-5}$ cm^2/s, $\alpha = 0.5$, and $T = 25$ °C. A spreadsheet calculates the parameters in equations (18.2.5) to (18.2.10) in columns D to I of Table 1 on page 160.

$$k_S = k^0 \exp[(1 - \alpha) f (E_S - E^{0\prime})] \qquad \text{(18.2.10, Col D)}$$

$$K_S = k_S a / D_R \qquad \text{(18.2.9, Col E)}$$

$$\Lambda = K_S d / a \qquad \text{(18.2.8, Col F)}$$

$$F(L, \Lambda) = \frac{\frac{11}{\Lambda} + 7.3}{110 - 40L} \qquad \text{(18.2.7, Col G)}$$

$$I_S = \frac{0.78377}{L(1 + \Lambda^{-1})} + \frac{0.68 + 0.3315 \exp\left(\frac{-1.0672}{L}\right)}{1 + F(L, \Lambda)} \qquad \text{(18.2.6, Col H)}$$

$$I_T(E_S, L) = I_S \left(1 - \frac{I_T^i}{I_T^c}\right) + I_T^i \qquad \text{(18.2.5, Col I)}$$

I_S and $I_T(E_S, L)$ are dimensionless. In column J of Table 1 (page 160), the current (A) is calculated as $i_{T,\infty} I_T(E_S, L)$. From equation (18.1.1), $i_{T,\infty} = GnFD_O C_o^* a$ and $RG = 10$ in Table 18.2.1 that yields geometric parameter $G = 4.07$. For conditions here, $i_{T,\infty} = 3.93 \times 10^{-9}$ A. The spreadsheet in Table 1 is shown for $k^0 = 10^{-2}$ cm/s, $E^{0\prime} = 0.000$ V and $C_0^* = 1.00$ mM. The initial substrate potential E_S is 1.00 V. These input constants are entered at the top of the spreadsheet as listed. The output voltammograms are shown for E_S varying from -1 to +1 V and $k^0 = 1 \times 10^{-2}$ cm/s.

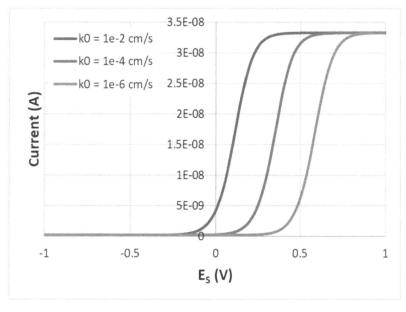

Input constants

a (cm) =	1.00E-03
C* (mol/cm^3) =	1.00E-06
L (no units) =	0.1
α (no units) =	0.5
D (cm^2/s) =	1.00E-05
F (coul/mol) =	96548
n (no units) =	1
f (V^{-1}) =	38.92
k^0 (cm/s) =	1.00E-02
E_S (V) =	1

The voltammograms simulated for $k^0 = 1 \times 10^{-2}$, 1×10^{-4}, and 1×10^{-6} are shown above. As k^0 decreases, the half wave potential shifts positive (away from the starting potential), which is consistent with slower electron transfer kinetics. The wave shapes are the same and all achieve the same limiting current. The sigmoidal shape is consistent with steady state diffusion.

Decreasing L (equivalent to decreasing d) results in a less reversible response, causing the i_T versus E_s curves to shift towards more positive potentials. This is due to the increased diffusive flux between tip and substrate at smaller d values. Limiting currents also decrease because flux to the electrode is more restricted.

	C	D	E	F	G	H	I	J	K

Table 1 shows values of i_T and intermediate parameters for $-1 < E_s < +1$ V and $k^0 = 10^{-2}$ cm/s

E_s (V)	k_s from (18.2.10)	K_s from (18.2.9)	Λ from (18.2.8)	$F(L, \Lambda)$ from (18.2.7)	I_s from (18.2.6)	$i_T(E_s,L)$ from (18.2.5)	$i_T(E_s,L)$
-1	3.54E-11	3.54E-09	3.54E-10	2.93E+08	5.09E-09	6.38E-02	2.51E-10
-0.95	9.36E-11	9.36E-09	9.36E-10	1.11E+08	1.35E-08	6.38E-02	2.51E-10
-0.9	2.48E-10	2.48E-08	2.48E-09	4.19E+07	3.56E-08	6.38E-02	2.51E-10
-0.85	6.55E-10	6.55E-08	6.55E-09	1.58E+07	9.43E-08	6.38E-02	2.51E-10
-0.8	1.73E-09	1.73E-07	1.73E-08	5.99E+06	2.49E-07	6.38E-02	2.51E-10
-0.75	4.59E-09	4.59E-07	4.59E-08	2.26E+06	6.60E-07	6.38E-02	2.51E-10
-0.7	1.21E-08	1.21E-06	1.21E-07	8.55E+05	1.75E-06	6.38E-02	2.51E-10
-0.65	3.21E-08	3.21E-06	3.21E-07	3.23E+05	4.62E-06	6.38E-02	2.51E-10
-0.6	8.50E-08	8.50E-06	8.50E-07	1.22E+05	1.22E-05	6.38E-02	2.51E-10
-0.55	2.25E-07	2.25E-05	2.25E-06	4.62E+04	3.23E-05	6.39E-02	2.51E-10
-0.5	5.95E-07	5.95E-05	5.95E-06	1.74E+04	8.56E-05	6.39E-02	2.51E-10
-0.45	1.57E-06	1.57E-04	1.57E-05	6.59E+03	2.26E-04	6.41E-02	2.52E-10
-0.4	4.16E-06	4.16E-04	4.16E-05	2.49E+03	5.99E-04	6.44E-02	2.53E-10
-0.35	1.10E-05	1.10E-03	1.10E-04	9.42E+02	1.58E-03	6.54E-02	2.57E-10
-0.3	2.91E-05	2.91E-03	2.91E-04	3.56E+02	4.19E-03	6.80E-02	2.67E-10
-0.25	7.71E-05	7.71E-03	7.71E-04	1.35E+02	1.11E-02	7.48E-02	2.94E-10
-0.2	2.04E-04	2.04E-02	2.04E-03	5.09E+01	2.91E-02	9.27E-02	3.64E-10
-0.15	5.40E-04	5.40E-02	5.40E-03	1.93E+01	7.56E-02	1.39E-01	5.46E-10
-0.1	1.43E-03	1.43E-01	1.43E-02	7.33E+00	1.92E-01	2.54E-01	1.00E-09
-0.05	3.78E-03	3.78E-01	3.78E-02	2.81E+00	4.64E-01	5.24E-01	2.06E-09
3.19189E-16	1.00E-02	1.00E+00	1.00E-01	1.11E+00	1.04E+00	1.09E+00	4.29E-09
0.05	2.65E-02	2.65E+00	2.65E-01	4.61E-01	2.11E+00	2.15E+00	8.46E-09
0.1	7.00E-02	7.00E+00	7.00E-01	2.17E-01	3.79E+00	3.82E+00	1.50E-08
0.15	1.85E-01	1.85E+01	1.85E+00	1.25E-01	5.69E+00	5.72E+00	2.25E-08
0.2	4.90E-01	4.90E+01	4.90E+00	9.00E-02	7.13E+00	7.14E+00	2.81E-08
0.25	1.30E+00	1.30E+02	1.30E+01	7.69E-02	7.91E+00	7.91E+00	3.11E-08
0.3	3.43E+00	3.43E+02	3.43E+01	7.19E-02	8.25E+00	8.25E+00	3.24E-08
0.35	9.08E+00	9.08E+02	9.08E+01	7.00E-02	8.39E+00	8.39E+00	3.30E-08
0.4	2.40E+01	2.40E+03	2.40E+02	6.93E-02	8.44E+00	8.44E+00	3.32E-08
0.45	6.36E+01	6.36E+03	6.36E+02	6.90E-02	8.46E+00	8.46E+00	3.33E-08
0.5	1.68E+02	1.68E+04	1.68E+03	6.89E-02	8.47E+00	8.47E+00	3.33E-08
0.55	4.45E+02	4.45E+04	4.45E+03	6.89E-02	8.47E+00	8.47E+00	3.33E-08
0.6	1.18E+03	1.18E+05	1.18E+04	6.89E-02	8.47E+00	8.47E+00	3.33E-08
0.65	3.11E+03	3.11E+05	3.11E+04	6.89E-02	8.47E+00	8.47E+00	3.33E-08
0.7	8.24E+03	8.24E+05	8.24E+04	6.89E-02	8.47E+00	8.47E+00	3.33E-08
0.75	2.18E+04	2.18E+06	2.18E+05	6.89E-02	8.47E+00	8.47E+00	3.33E-08
0.8	5.77E+04	5.77E+06	5.77E+05	6.89E-02	8.47E+00	8.47E+00	3.33E-08
0.85	1.53E+05	1.53E+07	1.53E+06	6.89E-02	8.47E+00	8.47E+00	3.33E-08
0.9	4.04E+05	4.04E+07	4.04E+06	6.89E-02	8.47E+00	8.47E+00	3.33E-08
0.95	1.07E+06	1.07E+08	1.07E+07	6.89E-02	8.47E+00	8.47E+00	3.33E-08
1	2.83E+06	2.83E+08	2.83E+07	6.89E-02	8.47E+00	8.47E+00	3.33E-08

19 SINGLE-PARTICLE ELECTROCHEMISTRY

Problem 19.2© For a 1 μm radius Pt disk electrode, where $D = 10^{-5}$ cm^{2}s^{-1}, $n = 1$, and $C^* = 1$ mM, the limiting current is expressed. a is the radius.

$$
\begin{aligned}
i_{\lim} &= 4nFDC^*a \\
&= 4\,(1)\left(96485\,\frac{C}{mol}\right)\left(10^{-5}\frac{cm^2}{s}\right)\left(10^{-6}\frac{mol}{cm^3}\right)\left(10^{-4}cm\right) \\
&= 3.8 \times 10^{-10}\ A
\end{aligned}
$$

The current response is sketched in the Figure.

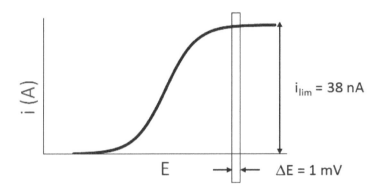

$i_{\lim}$ = 38 nA

ΔE = 1 mV

At 10 mV/s, the time Δt to scan 1 mV is 0.1 s. The number of moles of Fc, N_{Fc}, oxidized in $\Delta t = 0.1$ s is calculated from charge Q by Faraday's law.

$$
\begin{aligned}
N_{FC} &= \frac{Q}{nf} = \frac{i_{\lim}\Delta t}{nF} = \frac{\left(3.8 \times 10^{-10}\,\frac{C}{s}\right)(0.1\ s)}{(1)\left(96485\,\frac{C}{mol}\right)} \\
&= 3.9 \times 10^{-16}\ mol
\end{aligned}
$$

The number of Fc molecules oxidized in Δt is:

$$N_{Fc}\left(6.02 \times 10^{23}\ mol^{-1}\right) = \left(3.9 \times 10^{-16}\ mol\right)\left(6.02 \times 10^{23}\ mol^{-1}\right) = 2.3 \times 10^{8}\ Fc\ molecules$$

or 230 million molecules. The number is sufficiently large that the current is deterministic.

Even during a very small 1 μV interval of the SSV response (corresponding to 10^{-4} s), $\approx 10^5$ Fc molecules are oxidized. Thus, the instantaneous current represents the oxidation of a sufficiently large number of Fc molecules that the response can be accurately modeled by the differential equations that describe averaged local concentrations of Fc and Fc$^+$. Because time-dependent random fluctuations in the number of reactive Fc molecules at the electrode surface are simply too small, occur too fast relative to the experimental bandwidth, and are averaged out across the electrode surface, measurable electrical currents of steady-state measurements are generally not observed.

Electrochemical Methods: Fundamentals and Applications, Third Edition, Student Solutions Manual. Cynthia G. Zoski and Johna Leddy.
© 2025 John Wiley & Sons Ltd. Published 2025 by John Wiley & Sons Ltd.

Chapter 19 SINGLE-PARTICLE ELECTROCHEMISTRY

Problem 19.4©   The time t_d is defined as the average time for an ensemble of molecules to diffuse a fixed distance d. See section 19.6. The stated relationship between t_d and d is given by (19.6.1).

$$t_d = \frac{d^2}{2D} \tag{19.6.1}$$

Equation (4.4.3) relates the root-mean-squared (*rms*) displacement of molecules, $\bar{\Delta}$, to the continuous variable, t.

$$\bar{\Delta} = \sqrt{2Dt} \tag{4.4.3}$$

In both (19.6.1) and (4.4.3), D is the diffusion coefficient of the molecule. It is important to note that $\bar{\Delta}$ and d, as well as t and t_d, have different physical meanings, as noted above.

(4.4.3) is derived from a discrete random walk where a molecule steps a distance $\pm l$ during a time step τ. (See Figure 4.4.2 and Table 4.4.1.) Below we use an extended version of the discrete random walk shown in Figure 4.4.2 to demonstrate that (19.6.1) is correct.

A simple discrete random walk simulation can be readily performed using an Excel spreadsheet. This allows one to measure the time, $t_{d,j}$, for a single molecule j to diffuse a distance d. We note that because diffusion is a random process, the time it takes for each molecule to diffuse a distance d will be different. The Excel program provides a means to sequentially determine $t_{d,j}$ for N molecules. The average time for N molecules to diffuse a distance d is given by

$$t_d = \frac{1}{N} \sum_{j=1}^{N} t_{d,j} \tag{1}$$

The goal is to show that the value of t_d obtained from the simulation agrees with that predicted by the analytic expression (19.6.1).

A. Choosing values of d, l and τ for the Excel random-walk simulation

We note that both (19.6.1) and (4.4.3) are for one-dimensional diffusion. Thus, the Excel simulation is set up for a one-dimensional random walk. Here, we choose a value of $d = 100$ nm, as this is a typical gap distance between electrodes used in single-molecule redox cycling experiments.

As noted in the text (p. 194) directly below (19.6.1.) and in Footnote 8, D can be written in terms of l and τ.

$$D = \frac{l^2}{2\tau} \tag{2}$$

Let D be set to a typical value in aqueous solution for a small redox molecule, 10^{-5} cm^2/s. The value of l is then set to be a small fraction of the total distance, d (100 nm), in order to obtain an accurate simulation of diffusive transport. Here, we choose $l = 1$ nm (1% of d). τ is then computed from equation (2) to be 0.5 ns. Note that other values of l and τ can be chosen as long as the values satisfy equation (2). The Excel simulation presented below requires just a few seconds to compute the diffusion times of 200 molecules, so programming efficiency is not an issue for this problem.

B. Analytical calculation of t_d

Based on the above values of d (100 nm) and D (10^{-5} cm^2/s), t_d is computed from (19.6.1).

$$t_d = \frac{d^2}{2D} = \frac{\left(100 \text{ nm} \times 10^{-7} \frac{\text{cm}}{\text{nm}}\right)^2}{2 \times 10^{-5} \text{ cm}^2/\text{s}} = 0.5 \times 10^{-5} \text{ s} = 5 \ \mu s$$

The results of the random walk simulation for a large number of molecules (e.g., $N = 200$) should converge to this analytical value.

C. General approach to setting up random walk simulation in Excel for a single trajectory

For ease of notation in the simulation, the symbol t_{100} is used to represent the time for a single molecule to diffuse 100 nm. Note that t_{100} is equivalent to $t_{d,j}$ used in equation (1). The steps to set up the random walk simulation in Excel are as follows.

Initialize parameters.
1. Label B3 Transport Distance d (nm). Enter the value for d in C3. Here, 100.

2. Label B4 Diffusion Coefficient D (cm^2/s). Enter the value for D in C4. Here, 1E-5.

3. Label B6 Step Size l (nm). Enter the value for l in C6. Here, 1.

4. Label B7 Step Duration τ (ns). Compute the value for $\tau = l^2/2D$ in C7. Here, the answer is 0.5.

5. Label B9 Analytical t_{100} (μs). Compute the value for $t_{100} = d^2/2D$ in C9. Here, the answer is 5.

6. Label I4 t_{100} (ns). For a single trajectory j^{th} run of the simulation ($1 \leq j \leq N$), this finds the time for the jth run to first reach distance d.

Set up the random walk simulation.
1. Generate column of step numbers (Column G). In cell G7, enter 0. In cell G8, enter =G7+1. Drag G8 down for 10,000 steps to G10007.

2. Calculate the time corresponding to each step number (Column H). In cell H8, enter =G*C7. Drag H8 down.

3. Generate a random direction for each step (Column I). In cell I8, enter =IF(RANDBETWEEN(0,1)>0,1,-1). Drag I8 down.

4. Compute the current location for each step iteratively from the previous location and the direction of the next step (Column J). In cell J8, enter =IF((I8+J7)>0,(I8+J7),0). Drag J8 down.

5. Calculate t_{100} when the current distance is equal to the transport distance, d. In cell J4, enter =INDEX(H7:H10007, (MATCH(C3,J7:J10007,0))).

The table on page 164 shows results from one typical simulation. In this example, the molecule traveled 100 nm in 8,950 steps, corresponding to $t_{100} = 4.475$ μs. Each run of the simulation will be have some what similarly but t_{100} can vary widely.

The next steps are:
6. Repeat the calculation for 200 trajectories, compute the running average of t_{100}, and plot the running average against trial number.

7. Compute the histogram of t_{100} values based on 200 trajectories.

Representative Random Walk Calculation			
		t100 (ns)	4475
Step Number	Time (ns)	Direction	Current Location (nm)
0	0.0	1	0
1	0.5	1	1
2	1.0	1	2
3	1.5	-1	1
4	2.0	1	2
...	...	...	...
...	...	...	...
...	...	...	...
8946	4473.0	-1	98
8947	4473.5	1	99
8948	4474.0	-1	98
8949	4474.5	1	99
8950	4475.0	1	100

D. Output for N runs of the simulation

The simulation was repeated N times to obtain a statistically significant value of t_d. Here, $N = 200$ to generate samples for the statistics. Figure A shows t_{100} values for these 200 trials. Recall that each trial represents an individual molecule. It is clear that the arrival time t_{100} in each simulation is stochastic (i.e., random). However, the values of t_{100} cluster around the analytical value $t_d = 5$ μs, as expected (solid horizontal line).

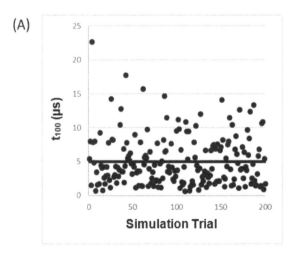

(A)

Figure B shows a plot of the running average of t_{100} for the 200 simulation trials. It is clear that the average value of t_{100} converges quickly to $t_d = 5$ μs. The average t_{100} from 200 trials is 4.97 μs, in excellent agreement with the analytical value of $t_d = 5$ μs. Thus, the results of the random walk simulations based on a discrete random walk (used to obtain (4.4.3)) are in agreement with (19.6.1). Closer agreement would be obtained by averaging t_{100} values over a larger number of trials (e.g., $N = 1000$).

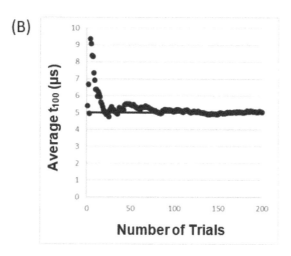

E. Histogram of Counts at each time from 0 to 25 μs for $N = 200$ runs of the simulation

Finally, Figure C and the accompanying table present the histogram of t_{100} values for the 200 simulations.

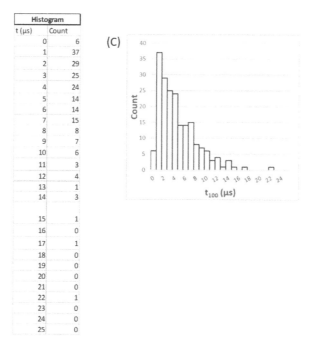

A smoother histogram would be obtained for larger N. It is interesting to note from these results that the shortest and longest values of t_{100} vary by well over an order of magnitude. Such is the nature of diffusion.

It is informative and amusing to examine histograms of arrival times (i.e., Figure C) as a function of distance, d. This is left to the student to explore by varying d in the spreadsheet.

Chapter 19 SINGLE-PARTICLE ELECTROCHEMISTRY

Problem 19.5[©] The dual thin layer system is shown in the figure. The electrodes are placed at $x = 0$ (cathode) and $x = d$ (anode). In the experiment, O is reduced at $x = 0$ and R is oxidized at $x = d$. To complete the circuit, the current at the two electrodes are equal. At steady state, the current at the two electrodes is equal and time invariant. The electrodes are separated by distance d and the electrode area is A

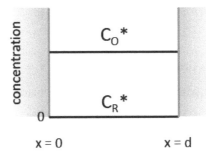

A single electron redox reaction $O + e \rightleftharpoons R$ is specified. Only O is initially present. The reaction is diffusion controlled. The diffusion coefficients for O and R are equal, $D_O = D_R = D$.

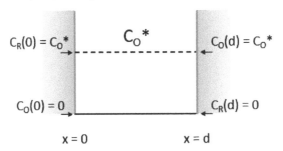

Initial conditions

The initial conditions are $c_O(x, 0) = C_O^*$ and $c_R(x, 0) = 0$ for $0 \leq x \leq d$.

Boundary conditions

The boundary conditions are set at the mass transport limit. All O that reaches the cathode at $x = 0$ is reduced to form R, so $c_O(0) = 0$ and $c_R(0) = C_O^*$. R diffuses from $x = 0$ toward $x = d$ and O diffuses from $x = d$ toward $x = 0$. At $x = d$ and the mass transport limit, all R that reaches the anode is oxidized to O. The surface concentrations are $c_R(d) = 0$ and $c_O(d) = C_O^*$.

In one dimension, Fick's second law defines the time and space dependence for the concentration of species i in time and space. The flux of i is $J_i(x, t)$.

$$\frac{\partial c_i(x, t)}{\partial t} = -\frac{\partial J_i(x, t)}{\partial x} = D \frac{\partial^2 c_i(x, t)}{\partial x^2}$$

At steady state, time invariance is specified as $\partial c_i(x, t) / \partial t = 0$. This requires $\partial J_i(x, t) / \partial x = 0$, which is only satisfied where $J_i(x)$ is a constant.

$$J_i(x) = -D\frac{dc_i(x)}{dx}$$

where both $J_i(x)$ and $c_i(x)$ are time invariant at steady state. Let constant $A' = -J_i(x)/D$.

$$A' = \frac{dc_i(x)}{dx}$$

A' is found by integration, where B' is the constant of integration for an indefinite integral.

$$\int dc_i(x) = \int A' dx$$
$$c_i(x) = A'x + B'$$

For O, the boundary conditions are $c_O(0) = 0$ and $c_O(d) = C_O^*$. Substitute the boundary conditions into $c_O(x) = A'x + B'$.

$$\text{At } x = 0 : c_O(0) = 0 = +B'$$
$$\text{At } x = d : c_O(d) = C_O^* = A'd + B' = A'd$$

Solve for $A' = \frac{C_O^*}{d}$. Then, $c_O(x)$ is defined.

$$c_O(x) = \frac{C_O^*}{d}x$$

$c_O(x)$ increases linearly with x. (Flux $J_O = -D\frac{C_O^*}{d}$ is constant for all x.)

For R, the boundary conditions are $c_R(0) = C_O^*$ and $c_R(d) = 0$. Substitute into $c_R(x) = A''x + B''$

$$\text{At } x = 0 : c_R(0) = C_O^* = B''$$
$$\text{At } x = d : c_R(d) = 0 = A''d + B'' = A''d + C_O^*$$

Solve for A'', $A'' = -\frac{C_O^*}{d}$. Then, $c_R(x)$ is defined.

$$c_R(x) = -\frac{C_O^*}{d}x + C_O^* = C_O^*\left(1 - \frac{x}{d}\right)$$
$$c_R(x) = \frac{C_O^*}{d}[d - x]$$

$c_R(x)$ decreases linearly with x. (Flux $J_R = -DA'' = D\frac{C_O^*}{d}$ is constant for all x.)

The linear concentration gradients $c_O(x)$ and $c_O(x)$ are shown on page 168. The flux of O and R are defined by the constant slopes. J_O and J_R are of equal magnitude but opposite signs.

Current is set by the concentration gradient at the electrode surface. The concentration gradient sets the flux. At steady state,

$$\frac{i_{ss}}{nFA} = -J_O = D\frac{dc_O(x)}{dx}\bigg|_{x=0} = D\frac{C_O^*}{d} \text{ for } O \tag{19.6.4}$$

If both O and R are initially present at C_O^* and C_R^*, then $C_T^* = C_O^* + C_R^*$. Once this system comes to steady state, the feedback establishes $i_{ss} = nFADC_T^*/d$.

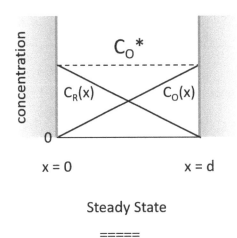

Steady State

=====

Problem 19.6© Twin electrode nanogap thin layer cell where Fc is at c_{bulk} =120 pM in the electrolyte that is fed to the cell through the ports. The cell dimensions are $L = 50$ μm; $w = 1.5$ μm; gap $d = 70$ nm.

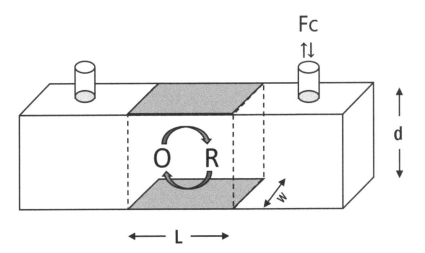

Redox cycling and generation of current occurs only in the solution volume where the electrode areas overlap. This defines the volume of the active region, Vol.

$$
\begin{aligned}
Vol &= d \times L \times w \\
&= \left(70 \times 10^{-7} \text{ cm}\right)\left(50 \times 10^{-4} \text{ cm}\right)\left(1.5 \times 10^{-4} \text{ cm}\right) \\
&= 5.25 \times 10^{-12} \text{ cm}^3 = 5.25 \times 10^{-15} \text{ L} = 5.25 \text{ pL}
\end{aligned}
$$

(a). The probability of 0, 1, 2, 3, $\cdots N$ Fc molecules between the electrodes is given by equation (19.3.6).

$$
P\left(N, \lambda\right) = \lambda^N \frac{\exp\left[-\lambda\right]}{N!} \tag{19.3.6}
$$

λ is the number of molecules in the volume averaged over time.

$$
\begin{aligned}
\lambda &= \langle N \rangle = c_{bulk} Vol N_A \\
&= \left(120 \times 10^{-12} \frac{\text{mol}}{\text{L}}\right) \left(5.25 \times 10^{-15}\ \text{L}\right) \left(6.02 \times 10^{23}\ \text{mol}^{-1}\right) \\
\lambda &= 0.3794\ \text{molecules} = \langle N \rangle
\end{aligned}
$$

Then, the probability $P(N, \lambda)$ of finding N molecules between the electrodes and the concentration $C(N)$ of N

$$
\begin{aligned}
P(N, \lambda) &= (0.3794)^N \frac{\exp[-0.3794]}{N!} = (0.3794)^N \frac{0.684}{N!} \\
C(N) &= \frac{N}{N_A Vol} = \frac{N}{3.161 \times 10^9} \frac{\text{mol}}{\text{cm}^3}
\end{aligned}
$$

The concentration $C(N)$ scales linearly with N.

N	0	1	2	3	4	5
$P(N, \lambda = 0.3794)$	0.684	0.260	0.0495	0.00622	5.9×10^{-4}	4.5×10^{-5}
$C(N)$ pm	0	316	632	948	1262	1580

The probability of finding no molecules in the volume is almost 70 %. The probability of one molecule in the volume is about 1 in 4. The probability of finding more than one molecule in the volume is 5.6 %.

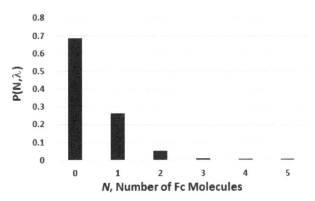

(b). The average concentration $\langle C \rangle$ in the volume is set by $P(N, \lambda)$ and $C(N)$.

$$
\begin{aligned}
\langle C \rangle &= \sum_i^N P(N, \lambda)\, C(N) \\
&= 0\,(0.684) + 1\,(0.260) + 2\,(0.0495) + 3\,(0.00622) + 4\,\left(5.9 \times 10^{-4}\right) + 5\,\left(4.5 \times 10^{-5}\right) \\
&= 0 + 82.8 + 31.1 + 5.89 + 0.75 + 0.071 \\
&= 120.0\ \text{pM}
\end{aligned}
$$

Chapter 19 SINGLE-PARTICLE ELECTROCHEMISTRY

As required by equilibrium, the concentrations in the electrolyte c_{bulk} and $\langle C \rangle$ are equal. For equilibrium, transport from the reservoir into the volume of the active region, Vol must be effective. In single particle measurements, where $d << w, L$, experiments balance transport with N and Vol.

(c). Substitution of equation (19.61), $t_d = d^2/2D$, into equation (19.6.4) leads to $i = nNe/(2t_d)$. On noting the concentration of the trapped species, $C^* = N/AdN_A$, this yields the macroscopic expression for steady state current in a thin layer, equation (19.6.6).

$$i = \frac{nFAD_eC^*}{d} \tag{19.6.6}$$

The reduced diffusion coefficient, $D_e = D$, when $D_O = D_R = D$. Assume $D_{Fc} = D_{Fc^+} = 1.7 \times 10^{-5}$ cm^2/s (Table C.4, for Fc in 0.1 M TBABF$_4$ in acetonitrile). For $Fc^+ + e \rightleftharpoons Fc$, $n = 1$. $C^* = 120.0 \times 10^{-12}$ mol/L $= 120.0 \times 10^{-15}$ mol/cm^3. $A = L \times w = \left(50 \times 10^{-4} \text{ cm}\right)\left(1.5 \times 10^{-4} \text{ cm}\right) = 7.5 \times 10^{-7}$ cm^2. $d = 70 \times 10^{-7}$ cm. The current is calculated.

$$
\begin{aligned}
i &= \frac{(1)\left(96485 \frac{\text{As}}{\text{mol}}\right)\left(7.5 \times 10^{-7} \text{ cm}^2\right)\left(1.7 \times 10^{-5} \frac{\text{cm}^2}{\text{s}}\right)\left(120.0 \times 10^{-15}\frac{\text{mol}}{\text{cm}^3}\right)}{70 \times 10^{-7} \text{ cm}} \\
&= 2.1 \times 10^{-14} \text{ A} = 21 \text{ fA}
\end{aligned}
$$

For the average concentration, the current is 21 fA. However, under conditions where a single ($N = 1$) molecule is trapped in the volume, $C(1) = 316$ pm. The single molecule concentration $C(1)$ higher than the average concentration $\langle C \rangle$ as 316 pm/120 pm $= 2.63$. When only one molecule is trapped, the current is calculated $i_{C(1)} = 2.63 \times i_{\langle C \rangle} = 2.63\,(21 \text{ fA}) = 55$ fA.

(d). Equation (16.6.4) describes current limited by transport of N molecules trapped in a thin layer in terms of the average time t_D for Brownian diffusion between the electrodes. From equation (19.6.1), $t_D = d^2/(2D) = \left(70 \times 10^{-7} \text{ cm}\right)^2 / \left(2 \times 1.7 \times 10^{-5} \text{ cm}^2/\text{s}\right) = 1.44 \times 10^{-6}$ s $= 1.4$ μs.

$$i = \frac{nNe}{2t_d} \tag{16.6.4}$$

$n = 1$, $N = 1$, and $e = F/N_A = (96485 \text{ C/mol})/(6.02 \times 10^{23} \text{ /mol}) = 1.60 \times 10^{-19}$ C per electron.

$$i = \frac{nNe}{2t_d} = \frac{(1)(1)\left(1.60 \times 10^{-19} \text{ As}\right)}{2\left(1.44 \times 10^{-6}\text{s}\right)} = 5.55 \times 10^{-14} \text{ A} = 56 \text{ fA}$$

The currents found from equations (19.6.6) and (16.6.4) are in agreement.

(e). One Fc molecule makes $\approx 10^6$ round trips across the cell in 1 second, spending 50 % of the time as Fc and 50 % as Fc$^+$. The time averaged position as Fc and Fc$^+$ would be identical to that predicted by Fick's 2nd law. That is, the time averaged behavior of 1 molecule over a long time is equivalent to the behavior of an ensemble of molecules. "Long time" is fairly short for this experiment because the cell dimensions are so small.

20 PHOTOELECTROCHEMISTRY AND ELECTROGENERATED CHEMILUMINESCENCE

Problem 20.1 [C] (a) An n-type semiconductor photovoltaic cell contains the redox couple

$$Fe(CN)_6^{3-} + e \rightleftharpoons Fe(CN)_6^{4-} \quad E^0 = 0.3610 \text{ V vs. NHE}$$

Measured with respect to SCE, $E^0 = 0.3610 \text{ V} -0.2412 \text{ V} = 0.1198 \text{ V}$ vs. SCE. With a flat-band potential, $E_{fb} = -0.20$ V vs. SCE, the $i - E$ curves appear as follows.

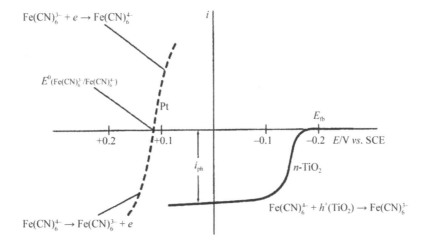

The open-circuit voltage under illumination is that voltage between the flat-band potential of the semiconductor electrode and the standard potential of the couple operating at the other electrode, or $E_{OC} = 0.1198 - (-0.20) = 0.32$ V.

The maximum short-circuit current under illumination is found from the light flux as follows:

$$i_{ph} = 6.2 \times 10^{15} \frac{\text{photons}}{\text{s}} \times \frac{e}{\text{photon}} \times \frac{1.602 \times 10^{-19} \text{ C}}{e} = 0.99 \text{ mA} \approx 1 \text{ mA}$$

(b) The expected output current vs. output voltage would have the trend shown below. Note that the difference between the maximum power output and the expected power output is due to the effects of the internal resistance of the cell and kinetic effects.

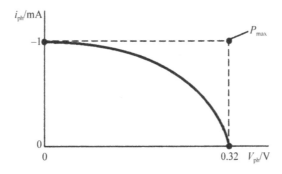

(c) The maximum output power for the cell is $P_{\text{max}} = i_{SC} \times V_{OC} = 1 \text{ mA} \times 0.32 \text{ V} = 0.32$ mW.

Electrochemical Methods: Fundamentals and Applications, Third Edition, Student Solutions Manual. Cynthia G. Zoski and Johna Leddy.
© 2025 John Wiley & Sons Ltd. Published 2025 by John Wiley & Sons Ltd.

Chapter 20   PHOTOELECTROCHEMISTRY AND ELECTROCHEMILUMINESCENCE

Problem 20.2$^{\copyright}$ (a) The thickness of the space charge region is given by equation (20.7.1).

$$L_1 = \sqrt{\frac{2\varepsilon\varepsilon_0}{eN_D}\Delta\phi} = \sqrt{1.1 \times 10^6 \varepsilon \frac{\Delta\phi}{N_D}} \text{ cm} \qquad (20.7.1)$$

where ε is the dielectric constant for the semiconductor; $\Delta\phi$ is the potential at the surface of the semiconductor with respect to the bulk semiconductor (V); and N_D is the donor density in number per cm^3. Let $\varepsilon = 10$. Below are calculaitons and plots of L_1 with $\Delta\phi$ for several values of N_D. On page 889, a doping level of 1 ppm is estimated at $\sim 5 \times 10^{16}$ cm^{-3}.

From equation (20.7.1) and the table, for a given N_D, L_1 increases as as $\Delta\phi$ increases. L_1 decreases as N_D increases. The magnitude of the curves shown for $L_1\,(\Delta\phi)$ diminishes as 10 ppb > 100 ppb > 1 ppm >10 ppm.

(b) When the band bending, $\Delta\phi = 0.5$ V, and the desired penetration depth for the light is 10^{-5} cm, then a good doping level is 1 ppm.

(c) At lower doping levels, the space charge thickness increases rapidly with $\Delta\phi$. The light will not be able to penetrate the semiconductor deeply enough to make use of most of the space charge region, where carrier separation can occur. At lower doping levels, the field in the space charge region is weaker; hence carrier separation may be less efficient. Also, the bulk resistance of the semiconductor rises which might impinge on the operating characteristics of the system.

	1 ppb	10 ppb	100 ppb	1 ppm	10 ppm	100 ppm
$N_D=$	5E+13	5E+14	5E+15	5E+16	5E+17	5E+18
$\Delta\phi$/V	L_1/cm	L_1/cm	L_1/cm	L_1/cm	L_1/cm	L_1/cm
0.00	0.00E+00	0.00E+00	0.00E+00	0.00E+00	0.00E+00	0.00E+00
0.05	1.05E-04	3.32E-05	1.05E-05	3.32E-06	1.05E-06	3.32E-07
0.10	1.48E-04	4.69E-05	1.48E-05	4.69E-06	1.48E-06	4.69E-07
0.15	1.82E-04	5.74E-05	1.82E-05	5.74E-06	1.82E-06	5.74E-07
0.20	2.10E-04	6.63E-05	2.10E-05	6.63E-06	2.10E-06	6.63E-07
0.30	2.57E-04	8.12E-05	2.57E-05	8.12E-06	2.57E-06	8.12E-07
0.40	2.97E-04	9.38E-05	2.97E-05	9.38E-06	2.97E-06	9.38E-07
0.50	3.32E-04	1.05E-04	3.32E-05	1.05E-05	3.32E-06	1.05E-06
0.60	3.63E-04	1.15E-04	3.63E-05	1.15E-05	3.63E-06	1.15E-06
0.70	3.92E-04	1.24E-04	3.92E-05	1.24E-05	3.92E-06	1.24E-06
0.80	4.20E-04	1.33E-04	4.20E-05	1.33E-05	4.20E-06	1.33E-06
0.90	4.45E-04	1.41E-04	4.45E-05	1.41E-05	4.45E-06	1.41E-06
1.00	4.69E-04	1.48E-04	4.69E-05	1.48E-05	4.69E-06	1.48E-06
1.10	4.92E-04	1.56E-04	4.92E-05	1.56E-05	4.92E-06	1.56E-06
1.20	5.14E-04	1.62E-04	5.14E-05	1.62E-05	5.14E-06	1.62E-06
1.30	5.35E-04	1.69E-04	5.35E-05	1.69E-05	5.35E-06	1.69E-06
1.40	5.55E-04	1.75E-04	5.55E-05	1.75E-05	5.55E-06	1.75E-06
1.50	5.74E-04	1.82E-04	5.74E-05	1.82E-05	5.74E-06	1.82E-06
1.60	5.93E-04	1.88E-04	5.93E-05	1.88E-05	5.93E-06	1.88E-06
1.70	6.12E-04	1.93E-04	6.12E-05	1.93E-05	6.12E-06	1.93E-06
1.80	6.29E-04	1.99E-04	6.29E-05	1.99E-05	6.29E-06	1.99E-06
1.90	6.47E-04	2.04E-04	6.47E-05	2.04E-05	6.47E-06	2.04E-06
2.00	6.63E-04	2.10E-04	6.63E-05	2.10E-05	6.63E-06	2.10E-06

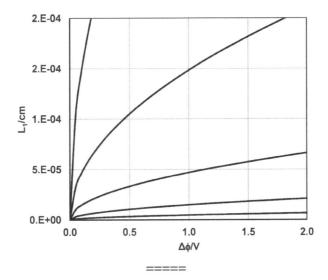

=====

Problem 20.3[©] The situation here is analogous to that in Figure 20.3.6. When the materials in Figure 20.7.1 are brought together, and the p-type semiconductor is taken to a negative potential relative to its flat-band potential, the bands near the surface of the semiconductor are bent upward toward the interior as shown below. Holes are drained toward the interior, so that a depletion layer develops at the surface.

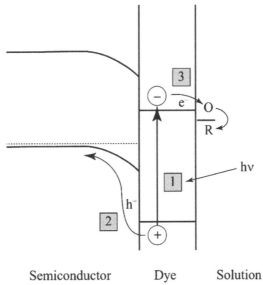

Semiconductor   Dye   Solution

In step 1, a photon is absorbed by the dye layer, creating an electron and a hole in the dye.

Because the filled states of the valence band of the semiconductor overlap with the valence band of the dye, the photogenerated hole in the dye layer is filled by electron transfer from the semi-conductor, leaving the hole in the semiconductor (step 2). the hole rises in the valence band of the semiconductor and is swept into the interior under the influence of the space-charge field.

In step 3, the remaining electron in the conduction band of the dye reduces species O in solution, producing species R, which can diffuse into the bulk.

Steps 2 and 3 are independent, so Step 3 could precede Step 2; however, the indicated order is far more likely because of the high density of filled states in the valence band of the semiconductor as compared to the density of empty states on species O in solution.

173

Chapter 20   PHOTOELECTROCHEMISTRY AND ELECTROCHEMILUMINESCENCE

The hole collected into the interior of the semiconductor can pass into an external circuit, then to a counter electrode such as Pt, where the hole can be used to oxidize species R (or another reductant) to preserve charge neutrality in the solution.

=====

Problem 20.4©   The Mott–Schottky equation is given in Section 20.2.1(d) of the text. For an n-type semiconductor electrode at 298 K,

$$\frac{1}{C_{SC}^2} = \frac{1.41 \times 10^{20}}{\varepsilon N_D} [-\Delta\phi - 0.0257] \tag{1}$$

where $-\Delta\phi = E - E_{fb}$. Or,

$$\frac{1}{C_{SC}^2} = \frac{1.41 \times 10^{20}}{\varepsilon N_D} [E - E_{fb} - 0.0257] \tag{20.2.3}$$

A plot of C_{SC}^{-2} versus E yields a slope of $m = 1.41 \times 10^{20}/\varepsilon N_D$. The intercept on the potential axis is $E_{fb} + 0.0257$.

For a p-type semiconductor, $\Delta\phi = E - E_{fb}$, and equation (1) becomes

$$\frac{1}{C_{SC}^2} = -\frac{1.41 \times 10^{20}}{\varepsilon N_A} [E - E_{fb} + 0.0257]$$

Accordingly, the slope of the Mott-Schottky plot is expected to be $-1.41 \times 10^{20}/\varepsilon N_A$, and the intercept on the potential axis should be $E_{fb} - 0.0257$ V.

For the data in Figure 20.2.3 recorded at 2500 Hz, slope and intercepts for the p-type and n-type data are as follows. The CRC Handbook gives the dielectric constant of InP as 12.4; this value was used to calculate N_D.

	p-type	n-type
slope (m⁴/F²V)	-3.63×10^4	37.8×10^4
intercept E axis (V)	0.88	-0.34
E_{fb} (V)	0.91	-0.37
εN_D (cm^{-3})	3.9×10^{15}	3.7×10^{14}
N_D (cm^{-3})	3.1×10^{14}	3.0×10^{13}

From these values, the difference in the flatband potentials for the n and p-type semiconductors is 1.28 V, which essentially matches the band gap of 1.3 eV for intrinsic InP.

For intrinsic InP, the electron and hole densities are equal, as approximated by equation (20.1.1).

$$
\begin{aligned}
n_i &= p_i \approx 2.5 \times 10^{19} \exp\left[\frac{-E_g}{2kT}\right] \text{ cm}^{-3} \text{ (near 25° C)} \tag{20.1.1}\\
&\approx 2.5 \times 10^{19} \exp\left[\frac{-1.3 \text{ eV}}{2 \times .02569 \text{ eV}}\right] \text{ cm}^{-3}\\
&\approx 2.57 \times 10^8 \text{ cm}^{-3}
\end{aligned}
$$

As expected, this is orders of magnitude smaller than the carrier level found for the doped InP.

A MATHEMATICAL METHODS

Problem A.1[©]   The definition of the Laplace transform is provided by equation (A.1.8).

$$L\left[F(t)\right] \equiv \int_0^\infty \exp\left[-st\right] F\left(t\right) dt \qquad (A.1.8)$$
$$= f\left(s\right)$$

The Laplace transform of $\sin at$ is found by evaluating the integral.

$$L\left[\sin(at)\right] \equiv \int_0^\infty \exp\left[-st\right] \sin(at) dt \qquad (1)$$

It is useful to note

$$\sin at = \frac{\exp\left[iat\right] - \exp\left[-iat\right]}{2i}$$

$$
\begin{aligned}
L\left[\sin(at)\right] &= \int_0^\infty \exp\left[-st\right] \frac{\exp\left[iat\right] - \exp\left[-iat\right]}{2i} dt \\
&= \frac{1}{2i} \int_0^\infty \exp\left[-st\right] \exp\left[iat\right] dt - \frac{1}{2i} \int_0^\infty \exp\left[-st\right] \exp\left[-iat\right] dt \\
&= \frac{1}{2i} \int_0^\infty \exp\left[-\left(s - ia\right) t\right] dt - \frac{1}{2i} \int_0^\infty \exp\left[-\left(s + ia\right) t\right] dt \\
&= \left(\frac{1}{2i}\right) \frac{\exp\left[-\left(s - ia\right) t\right]}{-\left(s - ia\right)} \bigg|_0^\infty - \left(\frac{1}{2i}\right) \frac{\exp\left[-\left(s + ia\right) t\right]}{-\left(s + ia\right)} \bigg|_0^\infty \qquad (2)
\end{aligned}
$$

In evaluating the upper limit of the integrations, it is noted that one requirement of the Laplace transform is that the transformed function be of exponential order, where $\sin at$ is of exponential order. (See Section A.1.2 in the text.) This means that there is a value of s where the argument of the integral in equation (A.1.8) is damped as $t \longrightarrow \infty$. Thus, the upper limits of the functions in equation (2) are zero.

$$
\begin{aligned}
L\left[\sin(at)\right] &= \frac{1}{2i} \frac{(0 - 1)}{-(s - ia)} - \frac{1}{2i} \frac{(0 - 1)}{-(s + ia)} \bigg| \\
&= \frac{1}{2i} \left(\frac{1}{s - ia} - \frac{1}{s + ia}\right) \\
&= \frac{1}{2i} \left(\frac{s + ia - (s - ia)}{s^2 + a^2}\right) \\
&= \frac{1}{2i} \left(\frac{+2ia}{s^2 + a^2}\right) = \frac{a}{s^2 + a^2}
\end{aligned}
$$

Electrochemical Methods: Fundamentals and Applications, Third Edition, Student Solutions Manual. Cynthia G. Zoski and Johna Leddy.
© 2025 John Wiley & Sons Ltd. Published 2025 by John Wiley & Sons Ltd.

Appendix A   MATHEMATICAL METHODS

Alternatively, equation (1) can be integrated by parts where

$$\int u \, dv = uv - v \, du$$

Here, it does not matter which function is set to u and which to dv. Let

$$
\begin{aligned}
u &= \exp[-st] & dv &= \sin at \, dt \\
du &= -s \exp[-st] & v &= -\cos[at]/a
\end{aligned}
$$

Then,

$$
\begin{aligned}
\int_0^\infty \exp[-st] \sin at \, dt &= \left. -\frac{\exp[-st] \cos at}{a} \right|_0^\infty - \frac{s}{a} \int_0^\infty \exp[-st] \cos at \, dt \\
&= \frac{1}{a}(0-1) - \frac{s}{a} \int_0^\infty \exp[-st] \cos at \, dt \\
&= \frac{1}{a} - \frac{s}{a} \int_0^\infty \exp[-st] \cos at \, dt
\end{aligned}
$$

Now, it is necessary to evaluate the integral on the right, again by parts. Let

$$
\begin{aligned}
u &= \exp[-st] & dv &= \cos at \, dt \\
du &= -s \exp[-st] & v &= \sin[at]/a
\end{aligned}
$$

Then,

$$
\begin{aligned}
\int_0^\infty \exp[-st] \sin at \, dt &= \frac{1}{a} - \frac{s}{a} \left\{ \left. \frac{\exp[-st] \sin[at]}{a} \right|_0^\infty + \frac{s}{a} \int_0^\infty \exp[-st] \sin at \, dt \right\} \\
&= \frac{1}{a} - \frac{s}{a^2} \times (0-0) - \frac{s^2}{a^2} \int_0^\infty \exp[-st] \sin at \, dt \\
&= \frac{1}{a} - \frac{s^2}{a^2} \int_0^\infty \exp[-st] \sin at \, dt
\end{aligned}
$$

Note the integral on the right and left are the same. This is rearranged as

$$
\begin{aligned}
\int_0^\infty \exp[-st] \sin at \, dt &= \frac{\frac{1}{a}}{1 + \frac{s^2}{a^2}} \\
&= \frac{a}{s^2 + a^2}
\end{aligned}
$$

This is the same the answer found by integration of the exponentials.

Problem A.5[©] **(a).** The Laplace transform of the problem is specified using equations (A.1.14) and (A.1.13).

$$\underbrace{s^2 y(s) - sY(0) - Y'(0)}_{L\{Y''(t)\}} + \underbrace{sy(s) - Y(0)}_{L\{Y'(t)\}} = 0$$

This rearranges to

$$\left(s^2 + s\right) y(s) - (s+1) Y(0) - Y'(0) = 0$$

Substitution of the boundary conditions $Y(0) = 5$ and $Y'(0) = -1$ yields

$$\left(s^2 + s\right) y(s) - 5(s+1) - (-1) = 0$$
$$\left(s^2 + s\right) y(s) - 5s - 4 = 0$$

Or

$$y(s) = \frac{5}{s+1} + \frac{4}{s^2 + s} \tag{1}$$

The inverse is taken termwise. For the first term, the inverse is found in Table A.1.1.

$$\frac{1}{s+a} \Leftrightarrow \exp[-at] \tag{2}$$

So,

$$\frac{5}{s+1} \Leftrightarrow 5\exp[-t]$$

The second term is not found directly in the Table. One option is to look in a more sophisticated table of transforms such as F. Oberhettinger and L. Badii, *Table of Laplace Transforms*, Springer-Verlag, New York, 1973, or even a general math table such as *CRC Mathematical Tables* from CRC Press or M. Abramowitz and I.A. Stegun (eds.), *Handbook of Mathematical Functions*, Dover Publications, New York. For example, the *CRC Mathematical Tables* lists

$$\frac{1}{(s-a)(s-b)} = \frac{1}{a-b} \left(\exp[at] - \exp[bt]\right)$$

For the second term, $b = 0$ and $a = -1$ such that

$$\frac{4}{s^2 + s} = \frac{4}{(s+1)s} = -4\left(\exp[-t] - 1\right) \tag{3}$$

Alternatively, equation (A.1.17) can be used to find the inverse of the second term. Note that $4/\left(s^2 + s\right) = (4/s)/(s+1)$. The inverse of $(s+1)^{-1}$ is given by equation (2). Equation (A.1.17) provides the method.

$$L\left\{\int_0^t F(x)\,dx\right\} = \frac{1}{s} f(s) \tag{A.1.17}$$

Thus,

$$L^{-1}\left\{\frac{1}{s} f(s)\right\} = \int_0^t F(x)\,dx$$

Or,

$$L^{-1}\left\{\frac{4}{s} \times \frac{1}{s+1}\right\} = 4\int_0^t \exp[-x]\,dx$$
$$= -4\exp[-x]\Big|_0^t$$
$$= -4\left\{\exp[-t] - 1\right\}$$

This is consistent with equation (3).

Appendix A   MATHEMATICAL METHODS

The inverse of equation (1) is now specified.

$$
\begin{aligned}
Y(t) &= 5\exp\left[-t\right] - 4\left[\exp\left(-t\right) - 1\right] \\
&= \exp\left[-t\right] + 4
\end{aligned}
$$

To verify that this is the correct solution,

$$
\begin{aligned}
Y'(t) &= -\exp\left[-t\right] \\
Y''(t) &= \exp\left[-t\right]
\end{aligned}
$$

such that

$$
Y''(t) + Y'(t) = \exp\left[-t\right] - \exp\left[-t\right] \overset{\checkmark}{=} 0
$$

and $Y(0) \overset{\checkmark}{=} 5$ and $Y'(0) \overset{\checkmark}{=} -1$.

(b). The Laplace transform of the problem is specified using the convolution or Faltung integral, equation (A.1.21), where τ is the variable of integration.

$$
\int_0^t F(t-\tau)G(\tau)d\tau = L^{-1}\left\{f(s)g(s)\right\} \tag{A.1.21}
$$

This can be re-expressed for the transform into Laplace coordinates as

$$
L\left\{\int_0^t F(t-\tau)G(\tau)d\tau\right\} = f(s)g(s)
$$

Thus,

$$
Y(t) = 2\cos t - 2\int_0^t Y(\tau)\sin(t-\tau)d\tau \tag{4}
$$

is transformed as

$$
y(s) = \frac{2s}{s^2+1} - 2y(s)\frac{1}{s^2+1}
$$

Solution for $y(s)$ yields

$$
y(s) = \frac{2s}{s^2+3}
$$

The inverse is found directly from Table A.1.1.

$$
Y(t) = 2\cos\sqrt{3}t
$$

To verify the solution, it will be necessary to evaluate the integral

$$
\int_0^t Y(\tau)\sin(t-\tau)d\tau = 2\int_0^t \cos\left(\sqrt{3}\tau\right)\sin(t-\tau)d\tau
$$

The exponential forms of $2i\sin x = \exp\left[ix\right] - \exp\left[-ix\right]$ and $2\cos x = \exp\left[ix\right] + \exp\left[-ix\right]$ should simplify the integration.

$$
\begin{aligned}
\cos\left(\sqrt{3}\tau\right)\sin(t-\tau) &= \frac{\exp\left[i\sqrt{3}\tau\right] + \exp\left[-i\sqrt{3}\tau\right]}{2} \\
&\times \frac{\exp\left[i\left(t-\tau\right)\right] - \exp\left[-i\left(t-\tau\right)\right]}{2i}
\end{aligned}
$$

$$
\begin{aligned}
4i\cos\left(\sqrt{3}\tau\right)\sin(t-\tau) &= \exp\left[it\right]\left\{\exp\left[i\left(\sqrt{3}-1\right)\tau\right] + \exp\left[-i\left(\sqrt{3}+1\right)\tau\right]\right\} \\
&- \exp\left[-it\right]\left\{\exp\left[i\left(\sqrt{3}+1\right)\tau\right] + \exp\left[-i\left(\sqrt{3}-1\right)\tau\right]\right\}
\end{aligned}
$$

Or,

$$\int_0^t Y(\tau)\sin(t-\tau)d\tau = \frac{2}{4i}\int_0^t \left\{ \exp[it]\left\{ \begin{array}{c} \exp\left[i\left(\sqrt{3}-1\right)\tau\right] \\ +\exp\left[-i\left(\sqrt{3}+1\right)\tau\right] \end{array}\right\} - \exp[-it]\left\{ \begin{array}{c} \exp\left[i\left(\sqrt{3}+1\right)\tau\right] \\ +\exp\left[-i\left(\sqrt{3}-1\right)\tau\right] \end{array}\right\} \right\}d\tau$$

$$= \frac{-i}{2}\exp[it]\int_0^t \left\{ \exp\left[i\left(\sqrt{3}-1\right)\tau\right] + \exp\left[-i\left(\sqrt{3}+1\right)\tau\right]\right\}d\tau$$

$$+\frac{i}{2}\exp[-it]\int_0^t \left\{ \exp\left[i\left(\sqrt{3}+1\right)\tau\right] + \exp\left[-i\left(\sqrt{3}-1\right)\tau\right]\right\}d\tau$$

$$= \frac{-i}{2}\exp[it]\left\{ \left.\frac{\exp\left[i\left(\sqrt{3}-1\right)\tau\right]}{i\left(\sqrt{3}-1\right)}\right|_0^t + \left.\frac{\exp\left[-i\left(\sqrt{3}+1\right)\tau\right]}{-i\left(\sqrt{3}+1\right)}\right|_0^t \right\}$$

$$+\frac{i}{2}\exp[-it]\left\{ \left.\frac{\exp\left[i\left(\sqrt{3}+1\right)\tau\right]}{i\left(\sqrt{3}+1\right)}\right|_0^t + \left.\frac{\exp\left[-i\left(\sqrt{3}-1\right)\tau\right]}{-i\left(\sqrt{3}-1\right)}\right|_0^t \right\}$$

$$= \frac{-i}{2}\exp[it]\left\{ \begin{array}{c} \frac{\exp[i(\sqrt{3}-1)t]}{i(\sqrt{3}-1)} - \frac{1}{i(\sqrt{3}-1)} \\ +\frac{\exp[-i(\sqrt{3}+1)t]}{-i(\sqrt{3}+1)} - \frac{1}{-i(\sqrt{3}+1)} \end{array}\right\}$$

$$+\frac{i}{2}\exp[-it]\left\{ \begin{array}{c} \frac{\exp[i(\sqrt{3}+1)t]}{i(\sqrt{3}+1)} - \frac{1}{i(\sqrt{3}+1)} \\ +\frac{\exp[-i(\sqrt{3}-1)t]}{-i(\sqrt{3}-1)} - \frac{1}{-i(\sqrt{3}-1)} \end{array}\right\}$$

$$= -\frac{1}{2}\left\{ \begin{array}{c} \frac{\exp[i\sqrt{3}t]}{\sqrt{3}-1} - \frac{\exp[it]}{\sqrt{3}-1} \\ -\frac{\exp[-i\sqrt{3}t]}{\sqrt{3}+1} + \frac{\exp[it]}{\sqrt{3}+1} \end{array}\right\} + \frac{1}{2}\left\{ \begin{array}{c} \frac{\exp[i\sqrt{3}t]}{\sqrt{3}+1} - \frac{\exp[-it]}{\sqrt{3}+1} \\ -\frac{\exp[-i\sqrt{3}t]}{\sqrt{3}-1} + \frac{\exp[-it]}{\sqrt{3}-1} \end{array}\right\}$$

$$= \frac{1}{2}\left\{ \begin{array}{c} \frac{\exp[i\sqrt{3}t]}{\sqrt{3}+1} - \frac{\exp[-it]}{\sqrt{3}+1} - \frac{\exp[-i\sqrt{3}t]}{\sqrt{3}-1} + \frac{\exp[-it]}{\sqrt{3}-1} \\ -\frac{\exp[i\sqrt{3}t]}{\sqrt{3}-1} + \frac{\exp[it]}{\sqrt{3}-1} + \frac{\exp[-i\sqrt{3}t]}{\sqrt{3}+1} - \frac{\exp[it]}{\sqrt{3}+1} \end{array}\right\}$$

$$= \frac{1}{2}\left\{ \begin{array}{c} \frac{\exp[i\sqrt{3}t]}{\sqrt{3}+1} + \frac{\exp[-i\sqrt{3}t]}{\sqrt{3}+1} - \frac{\exp[it]}{\sqrt{3}+1} - \frac{\exp[-it]}{\sqrt{3}+1} \\ +\frac{\exp[it]}{\sqrt{3}-1} + \frac{\exp[-it]}{\sqrt{3}-1} - \frac{\exp[i\sqrt{3}t]}{\sqrt{3}-1} - \frac{\exp[-i\sqrt{3}t]}{\sqrt{3}-1} \end{array}\right\}$$

$$= \frac{\cos\left[\sqrt{3}t\right] - \cos t}{\sqrt{3}+1} + \frac{\cos t - \cos\sqrt{3}t}{\sqrt{3}-1}$$

$$= \frac{1}{2}\left\{ \begin{array}{c} \left(\sqrt{3}-1\right)\cos\left[\sqrt{3}t\right] - \left(\sqrt{3}-1\right)\cos t \\ +\left(\sqrt{3}+1\right)\cos t - \left(\sqrt{3}+1\right)\cos\sqrt{3}t \end{array}\right\}$$

$$= \frac{1}{2}\left\{ -2\cos\left[\sqrt{3}t\right] + 2\cos t\right\}$$

$$= -\cos\left[\sqrt{3}t\right] + \cos t$$

Thus, equation (4) rearranges to the following, which verifies that $Y(t)$ is correct.

$$\int_0^t Y(\tau)\sin(t-\tau)d\tau = \frac{2\cos t - Y(t)}{2} = \frac{2\cos t - 2\cos\sqrt{3}t}{2} \overset{\checkmark}{=} \cos t - \cos\sqrt{3}t$$

Appendix A   MATHEMATICAL METHODS

(c). The Laplace transform is specified using equations (A.1.13) through (A.1.15) as

$$\underbrace{\frac{s}{s^2 + 1^2}}_{L\{\cos t\}} = \underbrace{s^3 y(s) - s^2 Y(0) - sY'(0) - Y''(0)}_{L\{Y'''(t)\}}$$

$$\underbrace{-\left[s^2 y(s) - sY(0) - Y'(0)\right]}_{L\{Y''(t)\}} \underbrace{-\left[sy(s) - Y(0)\right]}_{L\{Y'(t)\}} + \underbrace{y(s)}_{L\{Y(t)\}}$$

Combining terms and noting $Y(0) = Y'(0) = 0$ and $Y''(0) = 1$,

$$\frac{s}{s^2 + 1^2} = s^3 y(s) - 1 - s^2 y(s) - sy(s) + y(s)$$

$$= y(s)\left[s^3 - s^2 - s + 1\right] - 1$$

$y(s)$ is then

$$y(s) = \frac{1 + \frac{s}{s^2 + 1^2}}{s^3 - s^2 - s + 1}$$

$$= \frac{s^2 + s + 1}{\left(s^2 + 1^2\right)\left(s^3 - s^2 - s + 1\right)}$$

This will have to be inverted through partial fractions. Start by factoring the denominator. Note the terminal integer is 1 which suggests that possible divisors are $s - 1$, $s + 1$, $s - i$, and $s + i$. It is found that

and

$$s^3 - s^2 - s + 1 = (s + 1)(s - 1)^2$$

So,

$$s^2 + 1^2 = (s + i)(s - i)$$

$$\frac{s^2 + s + 1}{\left(s^2 + 1^2\right)\left(s^3 - s^2 - s + 1\right)} = \frac{s^2 + s + 1}{(s + i)(s - i)(s + 1)(s - 1)^2}$$

The rules for partial fractions are in Section A.1.4c in the text. The above becomes

$$\frac{s^2 + s + 1}{(s + i)(s - i)(s + 1)(s - 1)^2} = \frac{A}{s + i} + \frac{B}{s - i} + \frac{C}{s + 1} + \frac{D}{s - 1} + \frac{E}{(s - 1)^2} \tag{5}$$

Multiply both sides by the dominator to obtain

$$s^2 + s + 1 = A(s - i)(s + 1)(s - 1)^2 \tag{6}$$
$$+ B(s + i)(s + 1)(s - 1)^2$$
$$+ C(s + i)(s - i)(s - 1)^2$$
$$+ D(s + i)(s - i)(s + 1)(s - 1)$$
$$+ E(s + i)(s - i)(s + 1)$$

Let $s = 1$ and all the terms on the right hand side but E will be zero. What remains is

$$E = \frac{3}{(1 + i)(1 - i)(1 + 1)} = \frac{3}{4}$$

180

C is found by letting $s = -1$.

$$C = \frac{1}{(-1 + i)(-1 - i)(-1 - 1)^2} = \frac{1}{8}$$

B is found by letting $s = i$.

$$B = \frac{i}{(i + i)(i + 1)(i - 1)^2} = \frac{1 + i}{8}$$

A is found by letting $s = -i$.

$$A = \frac{-i}{(-i - i)(-i + 1)(-i - 1)^2} = \frac{1 - i}{8}$$

In evaluating A and B it is useful to note that expressions of the form $(a + bi)^{-1}$ are converted to expressions with the imaginary component in the numerator through the complex conjugate.

$$\frac{1}{a + bi} \times \frac{a - bi}{a - bi} = \frac{a - bi}{a^2 + b^2}$$

It remains to evaluate D. There is no choice of s that will make the other terms on the right hand side go to zero. A simple choice is to let $s = 0$. Then,

$$
\begin{aligned}
1 &= A(-i)(+1)(-1)^2 + B(+i)(+1)(-1)^2 + C(+i)(-i)(-1)^2 \\
&\quad + D(+i)(-i)(+1)(-1) + E(+i)(-i)(+1) \\
&= -iA + iB + C - D + E
\end{aligned}
$$

Substituting for the values of A, B, C, and E yields D.

$$
\begin{aligned}
D &= -iA + iB + C + E - 1 \\
&= -i\frac{1 - i}{8} + i\frac{1 + i}{8} + \frac{1}{8} + \frac{3}{4} - 1 \\
&= \frac{1}{8}[-1 - i(1 - i) + i(1 + i)] \\
&= -\frac{3}{8}
\end{aligned}
$$

Expanding equation (6) and substituting values for A, B, C, D, and E shows these values to be correct. Symbolic manipulators such as Maple, Mathematica, and Wolfram Alpha can efficiently reduce partial fraction problems.

Now equation (5) is expressed as

$$\frac{s^2 + s + 1}{(s + i)(s - i)(s + 1)(s - 1)^2} = \frac{(1 - i)/8}{s + i} + \frac{(1 + i)/8}{s - i} + \frac{1/8}{s + 1} - \frac{3/8}{s - 1} + \frac{3/4}{(s - 1)^2}$$

The inverse for the first four terms on the right arises through the inverse function in Table A.1.1.

$$\frac{1}{s + a} \Leftrightarrow \exp[-at]$$

The last term on the right hand side is not listed in Table A.1.1. It can be looked up in more extensive tables or it can be evaluated through the Shift Theorem, equation (A.1.18).

$$f(s - a) = L\{\exp[at] F(t)\} \tag{A.1.18}$$

Appendix A   MATHEMATICAL METHODS

From Table A.1.1,

$$\frac{1}{s^2} \Leftrightarrow t$$

Thus,

$$\frac{1}{(s-a)^2} \Leftrightarrow t \exp[at]$$

The inverse is now found as

$$
\begin{aligned}
Y(t) &= L^{-1}\left\{\frac{s^2+s+1}{(s+i)(s-i)(s+1)(s-1)^2}\right\}\\
&= L^{-1}\left\{\frac{(1-i)/8}{s+i}+\frac{(1+i)/8}{s-i}+\frac{1/8}{s+1}-\frac{3/8}{s-1}+\frac{3/4}{(s-1)^2}\right\}\\
&= \frac{1-i}{8}\exp[-it]+\frac{1+i}{8}\exp[it]+\frac{1}{8}\exp[-t]-\frac{3}{8}\exp[t]+\frac{3}{4}t\exp[t]\\
&= \frac{1}{8}\left\{\exp[-it]+\exp[it]+i(\exp[it]-\exp[-it])+\exp[-t]-3\exp[t]+6t\exp[t]\right\}\\
&= \frac{1}{8}\left\{2\cos t-2\sin t+\exp[-t]-3\exp[t]+6t\exp[t]\right\}
\end{aligned}
$$

To verify that this is the correct solution, find $Y'(t)$, $Y''(t)$ and $Y'''(t)$.

$$
\begin{aligned}
Y'(t) &= \frac{1}{8}\left\{-2\sin t-2\cos t-\exp[-t]-3\exp[t]+6\exp[t]+6t\exp[t]\right\}\\
&= \frac{1}{8}\left\{-2\sin t-2\cos t-\exp[-t]+3\exp[t]+6t\exp[t]\right\}
\end{aligned}
$$

$$
\begin{aligned}
Y''(t) &= \frac{1}{8}\left\{-2\cos t+2\sin t+\exp[-t]+3\exp[t]+6\exp[t]+6t\exp[t]\right\}\\
&= \frac{1}{8}\left\{-2\cos t+2\sin t+\exp[-t]+9\exp[t]+6t\exp[t]\right\}
\end{aligned}
$$

$$
\begin{aligned}
Y'''(t) &= \frac{1}{8}\left\{2\sin t+2\cos t-\exp[-t]+9\exp[t]+6\exp[t]+6t\exp[t]\right\}\\
&= \frac{1}{8}\left\{2\sin t+2\cos t-\exp[-t]+15\exp[t]+6t\exp[t]\right\}
\end{aligned}
$$

The original problem is now to be verified.

$$
\begin{aligned}
\cos t &\overset{?}{=} Y'''(t)-Y''(t)-Y'(t)+Y(t)\\
&\overset{?}{=} \frac{1}{8}\left\{2\sin t+2\cos t-\exp[-t]+15\exp[t]+6t\exp[t]\right\}\\
&\quad -\frac{1}{8}\left\{-2\cos t+2\sin t+\exp[-t]+9\exp[t]+6t\exp[t]\right\}\\
&\quad -\frac{1}{8}\left\{-2\sin t-2\cos t-\exp[-t]+3\exp[t]+6t\exp[t]\right\}\\
&\quad +\frac{1}{8}\left\{2\cos t-2\sin t+\exp[-t]-3\exp[t]+6t\exp[t]\right\}\\
&\overset{\checkmark}{=} \cos t
\end{aligned}
\tag{7}
$$

The other boundary conditions $Y(0)=Y'(0)=0$ and $Y''(0)=1$ are also satisfied.

Problem A.6[©] From Kirchoff's laws, the current i must equal the sum of the currents through the parallel components R and L in series with C. Let $i_1(t)$ be the current through the resistor and $i_2(t)$ be the current through the series capacitor and inductor.

$$i = i_1(t) + i_2(t)$$

Also, the voltage drops across R and LC must be equal. For the resistor,

$$V(t) = i_1(t)R \tag{1}$$

For the inductor in series with the capacitor,

$$V(t) = \frac{1}{C}\int_0^t i_2(\tau)d\tau + L\frac{di_2(t)}{dt}$$

Thus, combining these three equations yields

$$i_1(t)R = [i - i_2(t)]R = \frac{1}{C}\int_0^t i_2(\tau)d\tau + L\frac{di_2(t)}{dt}$$

$$i - i_2(t) = \frac{1}{RC}\int_0^t i_2(\tau)d\tau + \frac{L}{R}\frac{di_2(t)}{dt} \tag{2}$$

The Laplace transform of equation (2) is

$$\frac{i}{s} - \bar{i}_2(s) = \frac{1}{RC}\frac{\bar{i}_2(s)}{s} + \frac{L}{R}\left[s\bar{i}_2(s) - i_2(0)\right]$$

Initially, the current is zero ($i_2(0) = 0$) so this can be solved for $\bar{i}_2(s)$.

$$\bar{i}_2(s) = \frac{i/s}{\frac{1}{RCs} + \frac{L}{R}s + 1} = \frac{i}{s^2 + \frac{R}{L}s + \frac{1}{LC}} \tag{3}$$

As in the text on page 971, the inverse can be taken for an equation in the form of equation (A.1.30).

$$L\left[A\exp\left[-at\right]\sin bt\right] = \frac{Ab}{(s+a)^2 + b^2} \tag{A.1.30}$$

Equation (3) is not yet in this form, but it can be so expressed by completing the square. For $a = R(2L)^{-1}$, $(s+a)^2 = s^2 + Rs/L + R^2(2L)^{-2}$. Thus, $b^2 = (LC)^{-1} - R^2(2L)^{-2}$ such that $(s+a)^2 + b^2 = s^2 + Rs/L + (LC)^{-1}$. Then, $A = i/b$, where

$$b = \sqrt{\frac{1}{LC} - \frac{R^2}{4L^2}}$$

The inverse of equation (3) is then found.

$$i_2(t) = \frac{i}{\sqrt{\frac{1}{LC} - \frac{R^2}{4L^2}}}\exp\left[-\frac{Rt}{2L}\right]\sin\left[\sqrt{\frac{1}{LC} - \frac{R^2}{4L^2}}t\right]$$

and

$$i_1(t) = i - i_2(t) = i\left(1 - \frac{1}{\sqrt{\frac{1}{LC} - \frac{R^2}{4L^2}}}\exp\left[-\frac{Rt}{2L}\right]\sin\left[\sqrt{\frac{1}{LC} - \frac{R^2}{4L^2}}t\right]\right)$$

From equation (1), $V(t)$ is found.

$$V(t) = iR\left(1 - \frac{1}{\sqrt{\frac{1}{LC} - \frac{R^2}{4L^2}}}\exp\left[-\frac{Rt}{2L}\right]\sin\left[\sqrt{\frac{1}{LC} - \frac{R^2}{4L^2}}t\right]\right) \tag{4}$$

Appendix A   MATHEMATICAL METHODS

Note that as $t \rightarrow \infty$, the response reduces to $\lim_{t \rightarrow \infty} V(t) = iR$ as expected once the capacitor is charged.

In the spreadsheet, the response of the circuit to three different combinations of the components R, C, and L are calculated from equation (4). $V(t)/iR$ is plotted vs $RT/2L$ for $b = 0.1$, 1.0, and 10. The frequency of oscillations (ringing) increases with b.

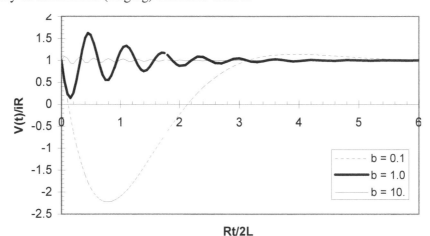

1/LC =	0.02	1.01	100.01
b =	0.1	1	10
R/2L =	0.1	0.1	0.1

Rt/2L	V(t)/iR	V(t)/iR	V(t)/iR
0	1	1	1
0.05	0.524583	0.543956	1.091216
0.1	0.09667	0.238606	1.049225
0.15	-0.286226	0.141448	0.944029
0.2	-0.626567	0.25553	0.925254
0.25	-0.926784	0.533909	1.010308
0.3	-1.189268	0.895456	1.073195
0.35	-1.41636	1.247193	1.030174
0.4	-1.610349	1.5073	0.950054
0.45	-1.773463	1.623301	0.945744
0.5	-1.907863	1.581617	1.015914
0.55	-2.015643	1.407061	1.057681
0.6	-2.098824	1.153346	1.016728
0.65	-2.15935	0.887698	0.956836
0.7	-2.19909	0.67375	0.96157
0.75	-2.219834	0.55692	1.018318
0.8	-2.223289	0.555453	1.044658
0.85	-2.211085	0.658715	1.007526
0.9	-2.18477	0.832445	0.963653
0.95	-2.145811	1.029064	0.973575
1	-2.095599	1.200134	1.018628
1.05	-2.035441	1.307839	1.033963
1.1	-1.966572	1.332868	1.001473
1.15	-1.890146	1.2772	0.970064
1.2	-1.807248	1.161613	0.982512
1.25	-1.718886	1.019002	1.01765
1.3	-1.626002	0.885491	1.025348
1.35	-1.529468	0.791627	0.997709
1.4	-1.430089	0.755719	0.975828
1.45	-1.32861	0.780701	0.989028
1.5	-1.225712	0.854901	1.015951
1.55	-1.122021	0.956178	1.018536
1.6	-1.018104	1.058127	0.99557
1.65	-0.914478	1.136698	0.980837
1.7	-0.811608	1.175631	0.993667
1.75	-0.709911	1.169538	1.013922

1/LC =	0.02	1.01	100.01
b =	0.1	1	10
R/2L =	0.1	0.1	0.1

Rt/2L	V(t)/iR	V(t)/iR	V(t)/iR
1.75	-0.709911	1.169538	1.013922
1.8	-0.609759	1.124137	1.013243
1.85	-0.511482	1.053851	0.99455
1.9	-0.415368	0.977583	0.985076
1.95	-0.321669	0.913847	0.996878
2	-0.2306	0.876446	1.011819
2.1	-0.057055	0.897546	0.994272
2.2	0.10416	1.000981	0.999021
2.3	0.252365	1.084841	1.006177
2.4	0.387234	1.082152	0.991423
2.5	0.508744	1.010864	1.007967
2.6	0.617119	0.943362	0.994925
2.7	0.712777	0.935726	1.001183
2.8	0.796293	0.983526	1.002358
2.9	0.868357	1.036515	0.99545
3	0.92974	1.049191	1.004977
3.1	0.981268	1.018202	0.996167
3.2	1.023795	0.977523	1.001745
3.3	1.058182	0.96312	1.000488
3.4	1.085282	0.982343	0.99783
3.5	1.105927	1.01293	1.002896
3.6	1.120913	1.027099	0.99738
3.7	1.130994	1.015911	1.001608
3.8	1.136877	0.99337	0.999704
3.9	1.139217	0.980491	0.999133
4	1.138613	0.986353	1.001559
4.2	1.130698	1.013744	1.00124
4.4	1.116831	0.999783	0.999784
4.6	1.099884	0.990935	0.999024
4.8	1.081982	1.006323	0.999493
5	1.064612	1.001768	1.000315
5.2	1.048736	0.994557	1.00055
5.4	1.034903	1.002524	1.000157
5.6	1.023343	1.001929	0.999736
5.8	1.014066	0.996994	0.999718
6	1.006926	1.000756	0.999989

Problem A.7[©] Taylor and Maclaurin series are useful for developing series approximations to functions that can be differentiated. If the series are developed about a value of interest, then good approximations about this value may be found for truncated series. Linear approximations are sufficient for many purposes.

(a). The Taylor series for a single independent variable is defined by equation (A.2.6).

$$f(x) = f(x_0) + \sum_{j=1}^{\infty} \frac{1}{j!} (x - x_0)^j \left[\frac{\partial^j}{\partial x^j} f(x) \right]_{x=x_0} \tag{A.2.6}$$

For a Taylor expansion of $\exp[ax]$ about $ax = 1$, let $ax = y$. This yields

$$
\begin{aligned}
\exp[y] &= \exp[1] + \sum_{j=1}^{\infty} \frac{1}{j!} (y-1)^j \left[\frac{\partial^j}{\partial y^j} \exp[y] \right]_{y=1} \\
&= \exp[1] + (y-1) \exp[y] \Big|_{y=1} + \frac{1}{2}(y-1)^2 \exp[y] \Big|_{y=1} + \frac{1}{3!}(y-1)^3 \exp[y] \Big|_{y=1} + \cdots \\
&= \exp[1] \left\{ 1 + (y-1) + \frac{1}{2!}(y-1)^2 + \frac{1}{3!}(y-1)^3 + \cdots \right\} = \exp[1] \sum_{j=0}^{\infty} \frac{1}{j!}(y-1)^j
\end{aligned}
$$

Thus, the linear approximation to the Taylor series about $y = ax = 1$ is

$$\exp[ax] = \exp[1][1 + (ax - 1)] = ax \exp[1] \tag{1}$$

(b). The Maclaurin series is defined in equation (A.2.7) and is appropriate about $x = 0$. It is a specific case of the Taylor expansion.

$$f(x) = f(0) + \sum_{j=1}^{\infty} \frac{1}{j!} x^j \left[\frac{\partial^j}{\partial x^j} f(x) \right]_{x=0} \tag{A.2.7}$$

For a Maclaurin expansion of $\exp[ax]$ about $ax = 1$, let $ax = y$. This yields

$$
\begin{aligned}
\exp[y] &= \exp[0] + \sum_{j=1}^{\infty} \frac{1}{j!} y^j \left[\frac{\partial^j}{\partial y^j} \exp[y] \right]_{y=0} \\
&= 1 + y \exp[y] \Big|_{y=0} + \frac{1}{2} y^2 \exp[y] \Big|_{y=0} + \frac{1}{3!} y^3 \exp[y] \Big|_{y=0} + \cdots \\
&= 1 + y + \frac{y^2}{2!} + \frac{y^3}{3!} + \cdots = \sum_{j=0}^{\infty} \frac{y^j}{j!}
\end{aligned}
$$

Thus, the linear approximation to the Maclaurin series about $y = ax$ is

$$\exp[ax] = 1 + ax \tag{2}$$

(c). Approximations that can be developed for the above Taylor and Maclaurin series would involve the number of terms needed to give an approximation of appropriate accuracy. The linear approximations would involve only the terms for $j = 0$ and $j = 1$. See equations (1) and (2). The plots below for the linear Taylor and Maclaurin approximations ($\times$) include the relative error (*) as compared to the real value of $\exp[y]$ (solid dark line). In each case, the approximations are within 2% for $y_0 \pm 0.2$. A better approximation is found for $j = 0$ to $j = 3$. These approximations ($\bigcirc$) and their relative error (light solid line) are plotted as well. In both cases, the relative error is below 2% for $y_0 - 0.7 \leq y \leq y_0 + 2.1$. Note that the two approximations give the same relative error about y_0, as anticipated for Taylor series and Maclaurin series because the Maclaurin is a Taylor developed about $y_0 = 0$.

| | Taylor Expansion | | | | | | | Maclaurin Expansion | | | | |
| | | | j=3 | j=3 | j=1 | j=1 | | | j=3 | j=3 | j=1 | j=1 |
y	exp y	y-1	app	% rel er	app	% rel er	y	exp y	app	% rel er	app	% rel er
0.0	1.000	-1.0	0.906	9.39	0.000	100.00	-1.0	0.368	0.333	9.39	0.000	100.00
0.1	1.105	-0.9	1.042	5.67	0.272	75.40	-0.9	0.407	0.384	5.67	0.100	75.40
0.2	1.221	-0.8	1.182	3.26	0.544	55.49	-0.8	0.449	0.435	3.26	0.200	55.49
0.3	1.350	-0.7	1.326	1.76	0.815	39.59	-0.7	0.497	0.488	1.76	0.300	39.59
0.4	1.492	-0.6	1.479	0.88	1.087	27.12	-0.6	0.549	0.544	0.88	0.400	27.12
0.5	1.649	-0.5	1.642	0.39	1.359	17.56	-0.5	0.607	0.604	0.39	0.500	17.56
0.6	1.822	-0.4	1.819	0.15	1.631	10.49	-0.4	0.670	0.669	0.15	0.600	10.49
0.7	2.014	-0.3	2.013	0.04	1.903	5.51	-0.3	0.741	0.741	0.04	0.700	5.51
0.8	2.226	-0.2	2.225	0.01	2.175	2.29	-0.2	0.819	0.819	0.01	0.800	2.29
0.9	2.460	-0.1	2.460	0.00	2.446	0.53	-0.1	0.905	0.905	0.00	0.900	0.53
1.0	2.718	0.0	2.718	0.00	2.718	0.00	0.0	1.000	1.000	0.00	1.000	0.00
1.1	3.004	0.1	3.004	0.00	2.990	0.47	0.1	1.105	1.105	0.00	1.100	0.47
1.2	3.320	0.2	3.320	0.01	3.262	1.75	0.2	1.221	1.221	0.01	1.200	1.75
1.3	3.669	0.3	3.668	0.03	3.534	3.69	0.3	1.350	1.350	0.03	1.300	3.69
1.4	4.055	0.4	4.052	0.08	3.806	6.16	0.4	1.492	1.491	0.08	1.400	6.16
1.5	4.482	0.5	4.474	0.18	4.077	9.02	0.5	1.649	1.646	0.18	1.500	9.02
1.6	4.953	0.6	4.936	0.34	4.349	12.19	0.6	1.822	1.816	0.34	1.600	12.19
1.7	5.474	0.7	5.442	0.58	4.621	15.58	0.7	2.014	2.002	0.58	1.700	15.58
1.8	6.050	0.8	5.995	0.91	4.893	19.12	0.8	2.226	2.205	0.91	1.800	19.12
1.9	6.686	0.9	6.596	1.35	5.165	22.75	0.9	2.460	2.427	1.35	1.900	22.75
2.0	7.389	1.0	7.249	1.90	5.437	26.42	1.0	2.718	2.667	1.90	2.000	26.42
2.1	8.166	1.1	7.956	2.57	5.708	30.10	1.1	3.004	2.927	2.57	2.100	30.10
2.2	9.025	1.2	8.720	3.38	5.980	33.74	1.2	3.320	3.208	3.38	2.200	33.74
2.3	9.974	1.3	9.544	4.31	6.252	37.32	1.3	3.669	3.511	4.31	2.300	37.32
2.4	11.023	1.4	10.431	5.37	6.524	40.82	1.4	4.055	3.837	5.37	2.400	40.82
2.5	12.182	1.5	11.383	6.56	6.796	44.22	1.5	4.482	4.188	6.56	2.500	44.22
2.6	13.464	1.6	12.403	7.88	7.068	47.51	1.6	4.953	4.563	7.88	2.600	47.51
2.7	14.880	1.7	13.493	9.32	7.339	50.68	1.7	5.474	4.964	9.32	2.700	50.68
2.8	16.445	1.8	14.657	10.87	7.611	53.72	1.8	6.050	5.392	10.87	2.800	53.72
2.9	18.174	1.9	15.897	12.53	7.883	56.63	1.9	6.686	5.848	12.53	2.900	56.63
3.0	20.086	2.0	17.216	14.29	8.155	59.40	2.0	7.389	6.333	14.29	3.000	59.40

Problem A.9© For a cell containing only electrolyte, the charge at the electrode surface can be viewed as a linear RC circuit established by a uncompensated resistance R_u and double layer capacitance C_d. The actual potential of the electrode E is defined relative to the potential of zero charge (PZC), E_z. At the PZC, the electrode is uncharged. The charge q at the electrode surface is then defined by the difference $E - E_z$, equation (1.6.11). If E is negative of E_z, the charge at the electrode surface is negative.

$$q = C_d (E - E_z) \tag{1.6.11}$$

At potential E_1 relative to E_z, the charge on the electrode surface is $q_1 = C_d (E_1 - E_z)$.

The cell is initially at equilibrium at potential E_1 where no current flows. At time $t = 0$, the applied potential, E_{appl}, is stepped to a value E_2. While E_{appl} changes effectively instantly from E_1 to E_2, the actual potential of the working electrode does not, because E_{appl} includes a voltage drop across the uncompensated resistance. The actual potential of the working electrode, E, is $E_{appl} + iR_u = E_2 + iR_u$ [equation (1.5.2)]. With the step, transient, nonfaradaic current i flows to charge the double layer. On substitution of $E = E_2 + iR_u$ into equation (1.6.11),

$$
\begin{aligned}
q &= C_d (E - E_z) = C_d (E_2 + iR_u - E_z) \\
&= C_d (E_2 - E_z) + iR_u C_d \\
q &= q_2 + iR_u C_d
\end{aligned}
\tag{1}
$$

Over time, i diminishes as the double layer becomes charged, and the actual potential approaches E_2, where the value of the charge is $q_2 = C_d (E_2 - E_z)$. Note that E, q, and i are time dependent and q_2, R_u, and C_d are constants. The time dependent q and i are related as

$$i = -\frac{dq}{dt} \tag{2}$$

On substitution into equation (1) and rearranging, a first order differential equation is found.

$$
\begin{aligned}
q &= q_2 - R_u C_d \frac{dq}{dt} \\
\frac{dq}{dt} &= \frac{q_2 - q}{R_u C_d} = -\frac{q}{R_u C_d} + \frac{q_2}{R_u C_d} \tag{1.6.14} \\
\frac{dq}{dt} &= aq + b \tag{3}
\end{aligned}
$$

where $a = -[R_u C_d]^{-1}$ and $b = q_2 [R_u C_d]^{-1}$.

The Laplace transform of equation 1.6.14 requires equation A.1.10.

$$L \left\{ \frac{dF(t)}{dt} \right\} = L \left[F'(t) \right] = s \bar{f}(s) - F(t = 0) \tag{A.1.10}$$

$F(t = 0)$ is the value of the function in the time domain. For the system here, the initial charge at

Appendix A   MATHEMATICAL METHODS

$t = 0$, $q_1 = C_d (E_1 - E_z)$ (equation 1.6.15). On Laplace transform,

$$L\left\{\frac{dq}{dt}\right\} = L\{aq + b\}$$

$$s\bar{q}(s) - q(t=0) = a\bar{q}(s) + \frac{b}{s}$$

$$s\bar{q}(s) - q_1 = a\bar{q}(s) + \frac{b}{s}$$

Or re-expressing with partial fractions,

$$\bar{q}(s)(s-a) = q_1 + \frac{b}{s}$$

$$\bar{q}(s) = \frac{sq_1 + b}{s(s-a)}$$

$$= \frac{A}{s} + \frac{B}{s-a}$$

Multiplication by $s(s-a)$ yields

$$sq_1 + b = A(s-a) + Bs$$

$$= s(A+B) - aA$$

$$sq_1 = s(A+B)$$

$$b = -aA$$

$$A = -\frac{b}{a} \quad B = q_1 - A = q_1 + \frac{b}{a}$$

For $b/a = -q_2$, the inverse yields

$$q = -\frac{b}{a} + \left(q_1 + \frac{b}{a}\right)\exp[at]$$

$$q = q_2 + (q_1 - q_2)\exp\left[-\frac{t}{R_u C_d}\right] \tag{4}$$

From equation (2),

$$i = -\frac{dq}{dt} = -\frac{d}{dt}\left\{q_2 + (q_1 - q_2)\exp\left[-\frac{t}{R_u C_d}\right]\right\}$$

$$i = \frac{q_1 - q_2}{R_u C_d}\exp\left[-\frac{t}{R_u C_d}\right] = \frac{E_1 - E_2}{R_u}\exp\left[-\frac{t}{R_u C_d}\right] = -\frac{\Delta E}{R_u}\exp\left[-\frac{t}{R_u C_d}\right]$$

ΔE is the potential step $E_2 - E_1$. For $q_1 - q_2 = -C_d\Delta E$, equation (4) is

$$q = q_2 - C_d\Delta E\exp\left[-\frac{t}{R_u C_d}\right] \tag{1.6.16}$$

The time constant $\tau = R_u C_d$. It is noted that for a potential step in a blank electrolyte (no redox species), $i(t)$ recorded for a potential step of magnitude ΔE can be plotted as $\ln i(t)$ versus t to yield R_u as ΔE normalized by the intercept. Given R_u, C_d is found from the slope $-[R_u C_d]^{-1}$. The current transient will have decayed to zero after about 4τ, at which point E will be E_2 and the electrode will be charged to q_2.

B BASIC CONCEPTS OF SIMULATIONS

Problem B.1[©] Consider equation (B.1.19), which characterizes the current under mass transport limited conditions.

$$Z(k+1) = \frac{i(k+1)t_k^{1/2}}{nFAD^{1/2}C^*} \tag{B.1.19}$$

This can be re-expressed at time t as

$$i(t) = Z(t)\frac{nFAD^{1/2}C^*}{t_k^{1/2}}$$

The Cottrell current arises following a potential step to the mass transport limit. It is given by equation (6.1.12).

$$i_d(t) = \frac{nFAD^{1/2}C^*}{\sqrt{\pi t}} \tag{6.1.12}$$

Then, for the Cottrell current at time t_k,

$$\begin{aligned}
\frac{i(t)}{i_d(t_k)} &= Z(t)\frac{nFAD^{1/2}C^*}{t_k^{1/2}} \times \frac{\sqrt{\pi t_k}}{nFAD^{1/2}C^*} \\
&= Z(t)\sqrt{\pi}
\end{aligned}$$

Or,

$$Z(t) = \frac{i(t)}{i_d(t_k)}\sqrt{\frac{1}{\pi}}$$

In this last equation, $Z(t)$ is proportional to ratio of currents with a proportionality factor of $\pi^{-1/2}$. As t approaches t_k, $i(t)$ approaches $i_d(tk)$, so Z approaches $\pi^{-1/2}$. You should be able to see this result in your simulations.

Electrochemical Methods: Fundamentals and Applications, Third Edition, Student Solutions Manual. Cynthia G. Zoski and Johna Leddy.
© 2025 John Wiley & Sons Ltd. Published 2025 by John Wiley & Sons Ltd.

Appendix B   BASIC CONCEPTS OF SIMULATIONS

Problem B.3[©]   Consider the current at time $k + 1$. The current expression for mass transport limited electrolysis is defined by equation (B.1.17). The general expression is similar.

$$i(k + 1) = \frac{nFADC^* [f_A(2, k) - f(1.k)]}{\Delta x}$$

Charge generated during a time step Δt is $q(k + 1)$ where

$$q(k + 1) = i(k + 1)\Delta t = \frac{nFADC^* [f_A(2, k) - f(1.k)] \Delta t}{\Delta x}$$

Given the dimensionless diffusion coefficient from equation (B.1.12), $\mathbf{D}_M = D\Delta t / \Delta x^2$, substitution for Δx yields the following:

$$\begin{aligned}
q(k + 1) &= \frac{nFADC^* [f_A(2, k) - f(1.k)] \Delta t \sqrt{\mathbf{D}_M}}{\sqrt{D\Delta t}} \\
&= nFA\sqrt{D}C^* [f_A(2, k) - f(1.k)] \sqrt{\Delta t}\sqrt{\mathbf{D}_M}
\end{aligned}$$

But, $\Delta t = t_k / \ell$.

$$q(k + 1) = nFA\sqrt{D}C^* [f_A(2, k) - f(1.k)] \sqrt{t_k}\sqrt{\mathbf{D}_M / \ell}$$

The usual process for defining a dimensionless parameter is to isolate all the dimensioned variables from the dimensionless terms. Let $Q(k + 1)$ be the dimensionless charge.

$$Q(k + 1) = \frac{q(k + 1)}{nFAC^*\sqrt{Dt_k}} = [f_A(2, k) - f(1.k)] \sqrt{\mathbf{D}_M / \ell}$$

From the units, $Q(k + 1)$ is dimensionless. The charge, like the current, is calculated at the next time $(k + 1)$ from the concentrations at the present time (k). From the discussion on page 992 in the text, the time to assign to the charge is better represented as $t/t_k = (k + 0.5)/\ell$ as opposed to $t/t_k = (k + 1)/\ell$. In part, this compensates for the forward difference used to derive the finite difference expression for the time derivative (see equation (B.1.6).

Problem B.4[©] Consider the reaction sequence

$$A + e \rightleftharpoons B \qquad \text{(at the electrode)}$$
$$B + C \overset{k_2}{\rightarrow} D \qquad \text{(in solution)}$$

The diffusion kinetic equation for species B combines Fick's law and the rate of consumption for B in the following reaction.

$$\frac{\partial C_B(x,t)}{\partial t} = D_B \frac{\partial^2 C_B(x,t)}{\partial x^2} - k_2 C_B(x,t) C_C(x,t)$$

This can be expressed in finite difference form based on equation (B.1.7) for the spatial second derivative and on equation (B.1.3) for the forward difference of the temporal first derivative.

$$\frac{C_B(x, t+\Delta t) - C_B(x,t)}{\Delta t} = D_B \left[\frac{C_B(x+\Delta x, t) - 2C_B(x,t) + C_B(x-\Delta x, t)}{\Delta x^2} \right]$$
$$-k_2 C_B(x,t) C_C(x,t)$$

Let $x = j\Delta x$ and $t = k\Delta t$. Then the above equation is expressed in indices of j and k as follows:

$$\frac{C_B(j, k+1) - C_B(j,k)}{\Delta t} = D_B \left[\frac{C_B(j+1, k) - 2C_B(j,k) + C_B(j-1, k)}{\Delta x^2} \right]$$
$$-k_2 C_B(j,k) C_C(j,k)$$

Solve for $C_B(j, k+1)$.

$$C_B(j, k+1) = C_B(j,k) + D_B \Delta t \left[\frac{C_B(j+1, k) - 2C_B(j,k) + C_B(j-1, k)}{\Delta x^2} \right]$$
$$-k_2 \Delta t C_B(j,k) C_C(j,k)$$

To generate fractional concentrations for B, normalize by the bulk concentration of B, C_A^*, such that $f_B(j,k) = C_B(j,k)/C_A^*$. To make the concentration of C fractional, normalize by C_C^* such that $f_C(j,k) = C_C(j,k)/C_C^*$.

$$f_B(j, k+1) = f_B(j,k) + D_B \Delta t \left[\frac{f_B(j+1, k) - 2f_B(j,k) + f_B(j-1, k)}{\Delta x^2} \right]$$
$$-k_2 C_C^* \Delta t f_B(j,k) f_C(j,k)$$

Appendix B   BASIC CONCEPTS OF SIMULATIONS

Note that $\mathbf{D}_{M,B} = D_B \Delta t / \Delta x^2$ and $\Delta t = t_k / \ell$.

$$
\begin{aligned}
f_B(j, k+1) &= f_B(j,k) + D_{M,B}\left[f_B(j+1,k) - 2f_B(j,k) + f_B(j-1,k)\right] \\
&\quad - \frac{k_2 C_C^* t_k}{\ell} f_B(j,k) f_C(j,k) \\
&= f_B'(j+1,k) - \frac{k_2 C_C^* t_k}{\ell} f_B(j,k) f_C(j,k)
\end{aligned}
$$

This yields an equation of the form of equations (B.3.11) and (B.3.12). The dimensionless rate constant is $k_2 C_C^* t_k / \ell$.

=====

Problems B.5 and B.6.- A Flux-Constrained Alternative

Spreadsheet simulations in Problems B.5 and B.6 are sensitive to input parameters. The current response can oscillate where flux to the electrode surface is insufficiently constrained. A flux-constrained alternative is provided by Steve Feldberg, in his first review of digital simulation in electrochemistry [S. W. Feldberg, *Electroanal. Chem.*, 3, 199 (1969), pp. 242-243]. Oscillations are suppressed because the dimensionless current is constrained by a measure of the flux at $x = 0$, so that the fractional concentrations ($f(j,k)$) remain positive and finite.